The Effect Evaluation of Haze Governance Policies in Hebei Province–Based on I-O Model

Guofeng Zhang · Xiaojing Ma · Huimin Li

The Effect Evaluation of Haze Governance Policies in Hebei Province–Based on I-O Model

Springer

Guofeng Zhang
School of Economic and Trade
Hebei GEO University
Shijiazhuang, Hebei, China

Xiaojing Ma
School of Resources
Hebei GEO University
Shijiazhuang, Hebei, China

Huimin Li
School of Economics and Management
Inner Mongolia University of Technology
Hohhot, Inner Mongolia, China

ISBN 978-981-15-3756-1 ISBN 978-981-15-3757-8 (eBook)
https://doi.org/10.1007/978-981-15-3757-8

This Springer imprint is published by the registered company Springer Nature Singapore Pte Ltd.
The registered company address is: 152 Beach Road, #21-01/04 Gateway East, Singapore 189721, Singapore

Preface

Along with the rapid social and economic development and the intensification of industrialization and urbanization, the pressure of energy and environmental carrying capacity in China has further intensified. The economic activities of human society are the main causes of fog and haze in China. The unreasonable structure of industry and energy utilization restricts the sustainable development of economy and environment in China.

Hebei province as a typical area, which plays a critical role in the process of fog and haze control in Beijing-Tianjin-Hebei region. The policies of fog and haze control should be further emphasized in Hebei province. A series of policies and measures have been introduced by Hebei province to control haze pollution, but the economic, social and environmental impact of these policies and measures is undiscovered. The purpose of this book is to find out the best way to realize the sustainable development of economy, society and environment in Hebei province by building models and simulating experiments. By comparing the socio-economic impacts in different scenarios, this book found the most effective policy combination of haze governance. Comprehensive haze governance policy recommendations provide experience for other regions of China and other developing countries. In this book, the dynamic simulation model of haze governance also provides a reference to other environmental policy simulations. Comprehensive haze governance policy recommendations provide experience for other regions of China and other developing countries. In this book, the dynamic simulation model of haze governance also provides a reference to other environmental policy simulations.

The book is divided into five parts:

The first part is an introduction. This part mainly introduces the research background, research status at home and abroad, the purpose and significance of the study, the content and methods of the study, the key scientific problems to be solved and the expected results. According to the national air quality data of 2016 published by the Ministry of Environmental Protection in February 2017, of the 74 cities monitored nationwide for air quality in 2016, Hebei province accounted for 6 of the 10 cities with the worst air quality. Haze control in Hebei province plays an important role in the process of that in Beijing-Tianjin-Hebei region (BTH region for short) and even in the whole country. In the process of joint prevention and control of air pollution in BTH region, the control focus should be inclined to Hebei. The comprehensive effects of various haze control schemes on economy, society and environment in Hebei province are still unclear. One of the core problems is that under the coupling relationship between pollutant emission and economic growth, it is urgent to find an optimal path to achieve rapid economic development and reduce haze pollution, so as to lay a foundation for long-term sustainable development.

In the second part, the current situation and existing problems of economic, social, energy and environment development in the study area are analyzed in detail. High-speed economic growth in Hebei province has put tremendous pressure on energy and environmental carrying capacity. The unreasonable structure of industry and energy utilization, and the rapid growth of motor vehicles have become the main reasons for the formation of fog and haze in Hebei province. From the perspective of comprehensive management of fog and haze in BTH region, The land area, resident population, economic output, total energy utilization and motor vehicle ownership of Hebei province accounted for 87%, 67%, 44%, 67% and 69% of those in the BTH region, respectively, which play a decisive role in the management of fog and haze in Beijing-Tianjin-Hebei region. In order to improve the air quality and promote the high-quality development of its economy, a series of policy documents have been issued by Hebei province, but the comprehensive impact of these policies on the economy, society and environment has not yet been discovered.

The third part constructs a comprehensive evaluation model of dynamic optimization of the haze control policies in Hebei province. The model includes an objective function, that is, maximizing economic and social development, and three sub-models including economic growth model, pollutant emission model and energy balance model. The year of 2012 is selected as the base period, and the haze control policy of Hebei province is written into the model as an exogenous variable, and the model is formulated and programmed. On the premise of removing the influence of natural factors on environmental quality, the model takes reducing the total emission of $PM_{2.5}$ as the environmental constraint target.

The simulation experiment is carried out, and the experimental results are analyzed in the fourth part. By comparing and analyzing the economic, social, energy and environmental indicators under different scenarios, the optimum scenario for realizing the sustainable development of economy, society, energy and environment in Hebei province is selected. In the optimum scenario, the sustainable development of the economy, the society, the energy and the environment of Hebei province can be achieved. First of all, the economy has grown and industrial structure have been optimized. From 2012 to 2030, the average annual growth rate of gross regional product (GRP) is 5.87%. By 2030, the proportion of added value of three industries will reach 13:37:50. Secondly, the energy utilization structure is optimized and the energy consumption intensity is reduced. By 2030, coal, oil, natural gas, traditional electricity and clean energy will account for 54.03%, 27.59%, 3.06%, 5.33% and 10% of the total energy utilization, respectively. In 2030, the energy utilization intensity will be 0.58 tons of standard coal/RMB 10,000 Yuan GRP, 49% less than that in 2012, and the energy utilization efficiency is greatly improved. Thirdly, the pollutants emissions have decreased annually, and the environmental and economic benefits have been gradually improved. Compared with 2016, the emissions of SO_2, NOx, PM_{10}, CO, VOCs, $PM_{2.5}$ and GHG intensity of Hebei province in 2030 will decrease by 52%, 40%, 19%, 18%, 6%, 32%, -3.3%, respectively.

The fifth part puts forward the policy suggestions for the realization of the economic, social, energy and environmental development of Hebei province. From the perspective of policy effect, the policies of subsidizing water and soil conservation, subsidizing the development and utilization of clean energy, subsidizing new energy vehicles, restricting the driving of motor vehicles and subsidizing the introduction of $PM_{2.5}$ treatment technology have diminishing effects. Compared to single policy, comprehensive policy can better achieve the goal of sustainable development of economy, society, energy and environment, and the effect of "source governance" policy is better than that of "end governance" policy. From 2017 to 2021, the comprehensive policy will achieve remarkable results in reducing emissions and effectively promote rapid economic development. In order to achieve sustained and rapid economic growth under the constraints of emission reduction targets, advanced policies and technologies need to be further introduced. In order to ensure that the predicted indicators of the economic, social, energy and environmental in the optimal scenario can be achieved, a total amount of 376.2 billion RMB Yuan is needed for subsidizing various environmental improvement policies from 2017 to 2030. The specific subsidies for the development and utilization of clean energy, the $PM_{2.5}$ dust removal technology, the soil and water conservation policies and the new energy vehicle policies are 335.38 billion RMB Yuan, 1.4 billion RMB Yuan, 3.4 billion RMB Yuan and 36.01 billion RMB Yuan, accounting for 89.15%, 0.37%, 0.91%, and 9.57% of the total subsidies, respectively.

This book is the research result of the Hebei Financial Support Project for the Introduced Overseas Student (CS201702) and phased research results of the National Social Science Fund (18BJY081).

Shijiazhuang, China — Guofeng Zhang
Shijiazhuang, China — Xiaojing Ma
Hohhot, China — Huimin Li

Contents

About the Authors

Guofeng Zhang male, born in 1981, Ph.D., Associate Professor, School of Economy and Trade, Hebei GEO University, research interests include environmental economics and resource economics.

Xiaojing Ma female, born in 1983, Ph.D., Lecturers, School of Resources, Hebei GEO University, research interests include geology, environmental economics and resource economics.

Huimin Li female, School of Economics and Management, Inner Mongolia University of Technology, research interests include environmental economics and finance.

Chapter 1
Introduction

1.1 Research Background and Research Status at Home and Abroad

1.1.1 Research Background

Human economic activities are closely related to air pollution. Rapid industrialization and urbanization are the main causes of air pollution worldwide (Akimoto 2003; Ru 2014). Serious haze pollution not only threatens human health, but also brings about huge economic losses (Word Bank 2007; WHO 2014; Gao 2015b; Wang 2016; Li 2017).

Over the past three decades, China has experienced rapid urbanization and economic development. According to the data released by the National Bureau of Statistics in January 2016, the urbanization rate in China had reached 56.1% by 2015. Rapid urbanization and industrialization are accompanied by such phenomena as population gathering, massive energy consumption, rapid increase in the number of automobiles, and increasing proportion of heavy industry, which lead to serious haze pollution (Yu 2014; Zhao 2018). According to the Asian Development Bank's 2012 report, less than 1% of China's 500 major cities met World Health Organization air quality standards (Asian Development Bank 2012). How to cope with the adverse effects of haze pollution on the economy, the society and the environment, and realize the sustainable development of economy, society and environment has become one of the critical problems to be solved urgently in China.

To this end, China promulgated and implemented the "Action Plan for the Prevention and Control of Atmospheric Pollution" in 2013, adopted a number of haze control programs to strive to achieve the goal of haze control from the source and the end. However, there are still disputes about the validity evaluation of the various schemes in the Plan (China Academy of Sciences Special Overall Group on "Tracing and Control of Atmospheric Haze" 2015; Jiang 2015; Ma 2019). The comprehensive effects of various haze control schemes on economy, society and environment

G. Zhang et al., *The Effect Evaluation of Haze Governance Policies in Hebei Province–Based on I-O Model*, https://doi.org/10.1007/978-981-15-3757-8_1

are still unclear. One of the core problems is that under the coupling relationship between pollutant emissions and economic growth, it is urgent to find an optimal path to achieve rapid economic development while reducing haze pollution, so as to lay the foundation for long-term sustainable development.

Hebei province is an essential province in BTH region. According to the data of National Statistical Yearbook 2015, the administrative land area of Hebei province is 188,800 km^2, accounting for 86.94% of the total area of BTH region, the population of Hebei province is 73,840,000, accounting for 67.08% of the total population of BTH region, the GRP of Hebei province in 2014 was 2.9 trillion RMB Yuan, accounting for 44.26% of the total GDP of BTH region, of which the industrial output value accounts for 55.18% of that in BTH region, the emission of SO_2 and NOx in 2014 of Hebei province was 1.19 million tons and 1.94 million tons, accounting for 81% and 78% of that in the BTH region, respectively. Coal combustion and industrial processes are the main sources of haze pollution in Hebei province. According to the national air quality data of 2016 published by the Ministry of Environmental Protection in January 2017, of the 74 cities monitored nationwide for air quality in 2016, six of the 10 cities with the worst air quality were from Hebei province. Therefore, the haze control in Hebei province plays a decisive role in the process of haze control in BTH region and in the whole country. In the process of implementing the joint prevention and control of air pollution in BTH region, the control should be focused on Hebei (China Academy of Sciences Special Overall Group on "Tracing and Control of Atmospheric Haze" 2015).

1.1.2 Research Status at Home and Abroad

1. **Understanding the formation process of haze pollution is the basis of implementing haze control scheme.**

At present, a preliminary consensus has been reached on the formation process of haze pollution. The sources of $PM_{2.5}$, the main pollutant of haze, can be divided into primary source (direct generation) and secondary source (secondary generation) (Bench 2004; Sun 2006; Kulkarni 2007; Tan 2009; Shen 2009; He Hong 2013; Liu 2014; Huang 2014). There are both natural and human causes in primary source (Mueller 2011; Song 2015a). Human causes include industrial dust, motor vehicle exhaust particulate matter, road dust, building dust, etc. Secondary generation is the secondary fine particles generated by gaseous pollutants (such as SO_2, NOx, NH_3, VOCs, etc.) discharged into the atmosphere through a variety of chemical and physical processes (Sun 2011; He 2013; Wang 2014; Huang 2014). Currently, $PM_{2.5}$ in China is mainly composed of secondary particles formed by the physical and chemical processes of the primary emission of pollutant gases, and less of them come from primary emission sources (Yang 2011; Quan 2011; Huang 2013; He Hong 2013; Huang 2014). Human economic activities are the main causes of haze.

2. **The causes of haze formation have regional differences**.

Low quality of urbanization and extensive economic growth mode are the fundamental causes of haze formation. The haze caused by the low quality of urbanization is mainly manifested as the haze pollution caused by the fast urbanization process, the rapid expansion of city scale, the expansion of population, the increasing building density, and the weakening of urban wind, dust, excessive energy consumption, automobile exhaust emissions caused by the increasing car ownership and other factors (Han 2014; Hu 2014; Tan 2018), where urbanization is advancing rapidly, haze pollution is more serious (Wu 2014; Hui 2018). The fog and haze caused by extensive economic growth mode is mainly manifested in the unreasonable structure of industry and energy. The extensive mode of economic growth has seriously distorted the industrial structure: the proportion of the primary industry which pollutes the environment is too high, the proportion of the output value of heavy industry with relatively heavy industrial pollution is increasing, while the proportion of the tertiary industry which pollutes the environment less is lagging behind (Zheng 2015). The more unreasonable the industrial structure, the greater the impact of urbanization on haze pollution (Zhao 2018). Extensive economic growth is also reflected in the unreasonable energy use structure, as of 2014, coal accounted for 66% of China's energy structure. The change of energy structure is highly correlated with haze pollution level, which shows that the proportion of coal in energy consumption varies in the same direction as haze pollution level. Due to the regional differences in resource endowment, economic development level and urbanization process, the formation mechanism of haze pollution also has regional differences (Choi 2013; Callen 2014; Cesari 2014; Hasheminassab 2014; Wu 2014; Du 2014; Li 2015; Wang 2015a, b, c; Chun 2016). Therefore, according to the formation mechanism of haze in different regions and the coupling relationship between its evolution and economic and social development, it is essential to scientifically plan urban development, change the mode of economic growth, adjust economic structure and energy consumption structure and supply structure to control haze (Sun 2018). However, there are few studies on the formation mechanism of fog and haze in Hebei province and the coupling relationship between its evolution and economy and society.

3. **"End-to-end" and "Source-to-source" are two critical measures for haze control.**

Haze pollution sources are generally divided into three categories: point source pollution including industrial production and residential life (excluding residential automobile energy consumption), non-point source pollution (haze pollution caused by different land use types), and mobile source pollution (traffic) (Cohan 2007; Hanna 2012; Zhuang 2014; Gao 2014; Guo 2014; Malla 2014; Wang 2015d; Zhang 2015). End-treatment is an important stage in the development of environmental management. Strengthening desulfurization, denitrification and dust removal in industrial production, raising emission standards of automobile exhaust and waste gas recovery and treatment are conducive to the rapid removal of haze pollution, and also to a certain extent slowing down the trend of environmental pollution and destruction caused by economic activities. However, end-treatment often does not involve the effective use of resources. It is not a thorough treatment, but the transfer of pollutants, and the operation cost is huge. Changing the mode of growth, promoting the upgrading of industrial structure, utilizing new energy and promoting clean technology are the fundamental measures to solve haze pollution from the source (Wang 2014; Zhuang 2014; Zhou 2015).

4. **Energy saving and emission reduction technology is the key to reduce haze pollution**.

Energy saving and emission reduction technologies include clean coal utilization (Shen 2015), renewable energy utilization (Lei 2014; Novan 2015), improving energy efficiency (Kanada 2013), clean fuel utilization (Bao 2014), boiler dust removal (He 2013) and environmental protection measures (such as VOCs control, building dust) in parks and construction industries (Chun 2016). From the specific types of pollutant removal technology, $PM_{2.5}$ emission reduction technologies mainly include inorganic membrane integrated multi-pollutant control and activated carbon coordinated multi-pollutant control technology, mixed particle collection and filter cleaning technology. With 99.95% removal rate of pollutants and export particle levels of the latest national emission standards, these technologies have been applied to industrial dust recovery and have achieved good economic and environmental benefits (Zhuang 2014). Catalytic technology, which reduces chemical substances such as nitrogen oxides (NH_3-SCR) or hydrocarbons (HC-SCR) by the catalytic technology, is the main method to remove NOx. It has been proved that catalytic technology can effectively remove nitrogen oxides (Wei 2013) from heavy diesel engines (He 2005; Liu 2008; Shan 2011) and coal combustion process (Wei 2013; Si 2014). SO_2 control technologies are mainly used to control sulfur dioxide flue gas of coal-fired boilers and sintering machines, including semi-dry internal and external circulating fluidized bed and wet desulfurization technology. So far, these two technologies have built demonstration projects in more than 30 areas of China, with an annual accumulation of 180,000 tons of desulfurization (Zhuang 2014). Environmental protection technology can effectively remove pollutants in the short term, but these measures cannot fundamentally control haze pollution. It is necessary to make comprehensive use of technologies and environmental economic policies from a systematic perspective in order to fundamentally solve the problem of haze pollution in the long term and achieve sustainable economic, social and environmental development.

5. **The management of fog and haze in Hebei province is exemplary**.

Hebei province is the province with the most haze polluted cities in China (Wang 2014). It plays a key role in the haze control process in China and BTH region. However, most of the existing studies focus on the overall environmental improvement effects of BTH region (Wang 2014; Chen 2015), ignoring the current economic and social development situation and resource endowment characteristics of Hebei province. Although improving emission standards and reducing industrial production can improve the environmental quality of BTH region (Huang 2015), it also inhibits the economic growth of Hebei province and is not conducive to the sustainable development of economy, society and environment of Hebei province. Therefore, it is necessary to build a comprehensive analysis model based on the characteristics of internal factors including economic, social, environmental and resource endowments and their characteristics of Hebei province, in addition, these internal factors should be considered together with the objectives of BTH regional environmental quality and the planning of governance (external factors) in order to achieve the objectives of environmental improvement in BTH region. In addition, the model can promote the economic transformation and development of Hebei province and provide a model for haze control in China.

6. **Based on the comprehensive evaluation of economic, social and environmental effects, the best way to control haze is explored.**

Urbanization and industrialization are the main topics of our country in the future. Each stage of urbanization and industrialization process has different characteristics. With the development of urbanization and economy, the causes and mechanisms of haze formation also change, and the haze control scheme needs to be adjusted accordingly. A key scientific problem is how to synthetically evaluate the economic and environmental effects of various schemes of "source control" and "end control" according to the coupling relationship between the formation cause, mechanism and evolution process of haze and economic and social development in a specific region, and explore the optimal path of haze control. However, the existing research is not enough to answer the above questions. Currently, studies on the impact of haze control schemes focus on "post assessment" (Wang 2014; Zheng 2015; Yuan 2015), or only on cost and benefit (Muller 2012; Chiesa 2014; Gao 2015), air quality improvement and direct economic benefits of reducing health damage caused by a single policy or technology reduction (Hixson 2009; Zhang 2010; Hanna 2012; Rao 2013; Vedrenne 2015; Qiu 2015; Jiang 2015). There is insufficient attention paid to the comprehensive evaluation and prediction of haze control schemes.

Therefore, the optimal path planning of haze control in the process of urbanization and industrialization needs a dynamic and comprehensive model, which can simultaneously reveal the economic and environmental effects of technological progress and economic transformation. At present, there are two main methods to construct the environmental economic evaluation model of haze control: Bottom-up model and Top-down model. From the engineering point of view, Bottom-up model chooses energy technology with cost advantage by predicting the change of technology and cost structure caused by technological innovation or new energy use. It has high credibility in evaluating the substitution effect of resource production technology, such as Markal model (Rafaj 2007), LEAP model (Mao 2014), and EDA model (Sueyoshi 2015), etc. However, this kind of model ignores the feedback of general economic and non-technical market factors. Top-down model analyses the impact of economic development on various sectors from the macroeconomic point of view. At the top of the model, production function is used to connect input of production factors with economic output. The production function is set to reflect the substitution relationship between production factors. Analytical techniques and policies are used to analyze the impact of economic growth and environmental change, such as CGE model (Rive 2010; Chen 2014), system dynamics model (Vafa-Arani 2014), etc. The disadvantage of this kind of modeling method is that the description of resource production and utilization technology is abstract, and the reason for the change of resource consumption is not clear enough.

7. **The dynamic optimization comprehensive evaluation model**.

The dynamic optimization comprehensive evaluation model, which combines dynamic linear programming with input-output model, can organically integrate the "top-down" macro-environmental economic evaluation model (based on input-output model) with the "bottom-up" technology evaluation model (Xiang 2014). The characteristic of this kind of model is that it is not limited by the scale and variables of the model, and can integrate multiple schemes into the model for comprehensive evaluation. In the dynamic optimization model, not only can economic growth be set as the objective function, resource utilization efficiency and environmental pollution level as constraints, the optimal economic development path can be discussed, but also environmental pollution minimization can be set as the objective function, economic growth rate and resource utilization efficiency as constraints, and thus the optimal environmental improvement path can be explored. In the meanwhile, according to the external environment, environmental change or development goals need to be further set as more specific and detailed objective functions and constraints (Zhang 2014).

1.2 Purpose and Significance of the Research

1.2.1 Research Purpose

This book has the following two research objectives:

(1) By constructing a dynamic optimization model of haze control path, this book aims to provide a model framework for the sustainable development of regional economy, society and environment.
(2) Taking Hebei province as an example, this book makes an empirical analysis and explores the best way to control haze in the process of urbanization and industrialization, so as to provide a scientific basis for the sustainable development of economy, society and environment in Hebei province.

1.2.2 Research Significance

The research significance of this book lies in the following two points:

(1) Based on the current situation of economy, society and environment in Hebei province, the causes, mechanism and evolution process of haze formation in Hebei province are analyzed, and the coupling relationship between economic and social development and the evolution of haze formation mechanism is studied. On this basis, the dynamic optimization model of haze control path is constructed, which enriches the theoretical basis of haze control.
(2) Based on the requirements of regional sustainable development, different scenarios are designed to compare and analyze the economic, social and environmental effects of various haze control schemes, which provides a scientific experimental basis for finding out the optimal path of economic, social and environmentally sustainable development in Hebei province, and has high application value.

1.3 Research Content, Research Methods and Technical Route

1.3.1 Research Content

This book mainly includes three parts:

(1) Study on the formation reasons and evolution mechanism of fog and haze in Hebei province

This Book selects Hebei province as the empirical study. The population, economic aggregate and economic structure, energy supply and consumption structure, sources of atmospheric pollutants, emission structure and emission factors of Hebei province since 2000 is investigated and studied, and regression analysis is conducted to analyze the causes, mechanism and evolution process of haze formation in Hebei province. Based on the coupling relationship between haze pollution and economic development, a dynamic optimization model was constructed to simulate the haze control path.

(2) Constructing a dynamic optimization model for simulating haze control

The variables of the dynamic optimization model consist of endogenous variables and exogenous variables. The constraints are the planning objectives of economic and social development and environmental improvement. The model structure includes an objective function (set according to specific research needs, such as maximizing economic growth or minimizing environmental pollution) and three sub-models (economic and social module, environmental module and energy module).

The economic and social system uses dynamic input-output model to describe the impact of haze control scheme on capital input and social output, and the relationship among various sectors within the economy. Specifically, the horizontal balance of input and output table is used to simulate the coupling relationship between technological progress, industrial investment, consumption and net export and total output. In addition, the vertical balance of the input-output table is used to simulate the coupling relationship between energy tax and total input.

The environmental system describes the emission of air pollutants from economic activities. $PM_{2.5}$ is selected as the measurement index (including primary and secondary sources) in tons. Pollution sources include point sources (industry and residents), non-point sources (land use) and mobile sources (transportation).

Energy system describes the balance between energy supply and demand in economic activities. Energy supply includes traditional petrochemical energy and new energy. Energy demand includes industrial production, residents' life and transportation.

In this dynamic optimization model, the three sub-models are interdependent and mutually restrictive. Economic activities consume energy, resources and other means of production, and discharge air pollutants into the environment. The government subsidizes haze pollution control programs (including the introduction of advanced technology) to achieve the goal of economic development and environmental improvement. The calculation of different variables among the models is transformed with corresponding coefficients to unify the dimension. After the model is built, the computer language is written and the base period data is entered, and trial calculations are performed using Lingo software to measure the stability and sensitivity of the coefficients in the model, debug the model, and finally build a dynamic optimization model.

(3) Simulated experimental study on optimum path of haze control

Using the data of Hebei province in 2012 as the base period data and Lingo software as the simulation experiment, the simulation period is 2012–2030, in which 2012–2015 is the simulation calibration period. It is necessary to calibrate the external parameters based on literature and practical research, so that the model results are consistent with the actual changes of the period, and 2016–2030 is the simulation period.

According to different economic development and environmental improvement objectives, different scenarios are set to conduct simulation experiments, and the economic growth rate, industrial structure adjustment trend and regional industrial layout, total energy supply and consumption and structural change trend, total pollutant emission reduction and environmental improvement, policy arrangement and technology selection and financial input in various scenarios are forecasted. Finally, by comparing various economic and environmental indicators in different scenarios, the optimal path to achieve sustainable economic, social and environmental development is found out, which provides a scientific basis for achieving sustainable economic, social and environmental development.

1.3.2 Research Program and Technical Route

The technical route adopted in this book is shown in Fig. 1.1. Firstly, the research framework is established. Secondly, by using the methods of literature research and field investigation, the relevant data of economic and social development and air pollution in Hebei province, as well as the planning objectives and treatment schemes of current treatment are obtained, the sources of pollution and pollutant emission factors are determined. The data are statistically analyzed to explore the causes, mechanism and evolution process of haze control in Hebei province, and dynamically fit the situation of economic society and haze in Hebei province. Thirdly, based on the above data and analysis results and existing research foundation, the dynamic optimization model of haze control path in Hebei province is constructed, and the model is translated into computer language by Lingo software. Finally, the comprehensive impact of haze control schemes on economy, society and environment is predicted and evaluated by using simulation experiment method. The optimal path of haze control in the process of urbanization and industrialization is discussed through analysis and comparison of experimental results, which provides scientific experimental basis for the sustainable development of regional economy, society and environment.

1.4 Key Scientific Questions to be Addressed and Expected Accomplishments

(1) Based on the process of economic and social development and environmental change in Hebei province, this book explores the correlation and uniqueness of the causes and evolution mechanism of haze in Hebei province. The emission sources of $PM_{2.5}$, SO_2, NOx and VOCs from primary sources are identified. The emission factors of various pollutants and the secondary generation coefficients of $PM_{2.5}$ from SO_2, NOx and VOCs are determined. The coupling relationship between haze pollution and economic activities was fitted.

(2) This book discusses the general model framework and regional difference of haze governance path simulation. By studying the planning objectives and available treatment schemes of regional haze control, setting model objective functions and constraints, and selecting endogenous and exogenous variables, the stability and sensitivity of coefficients in the model are determined through data trial calculation, and the model is further debugged to verify its scientificity, rationality and operability. Finally, a dynamic and comprehensive model with regional characteristics is constructed.

(3) Route optimization simulation and result evaluation of haze control in Hebei province. By comparing the simulation results and analyzing the indicators of economy, society and environment under various scenarios, the optimal policy and technology combination for realizing the sustainable development of economy, society and environment in Hebei province is found out, and the

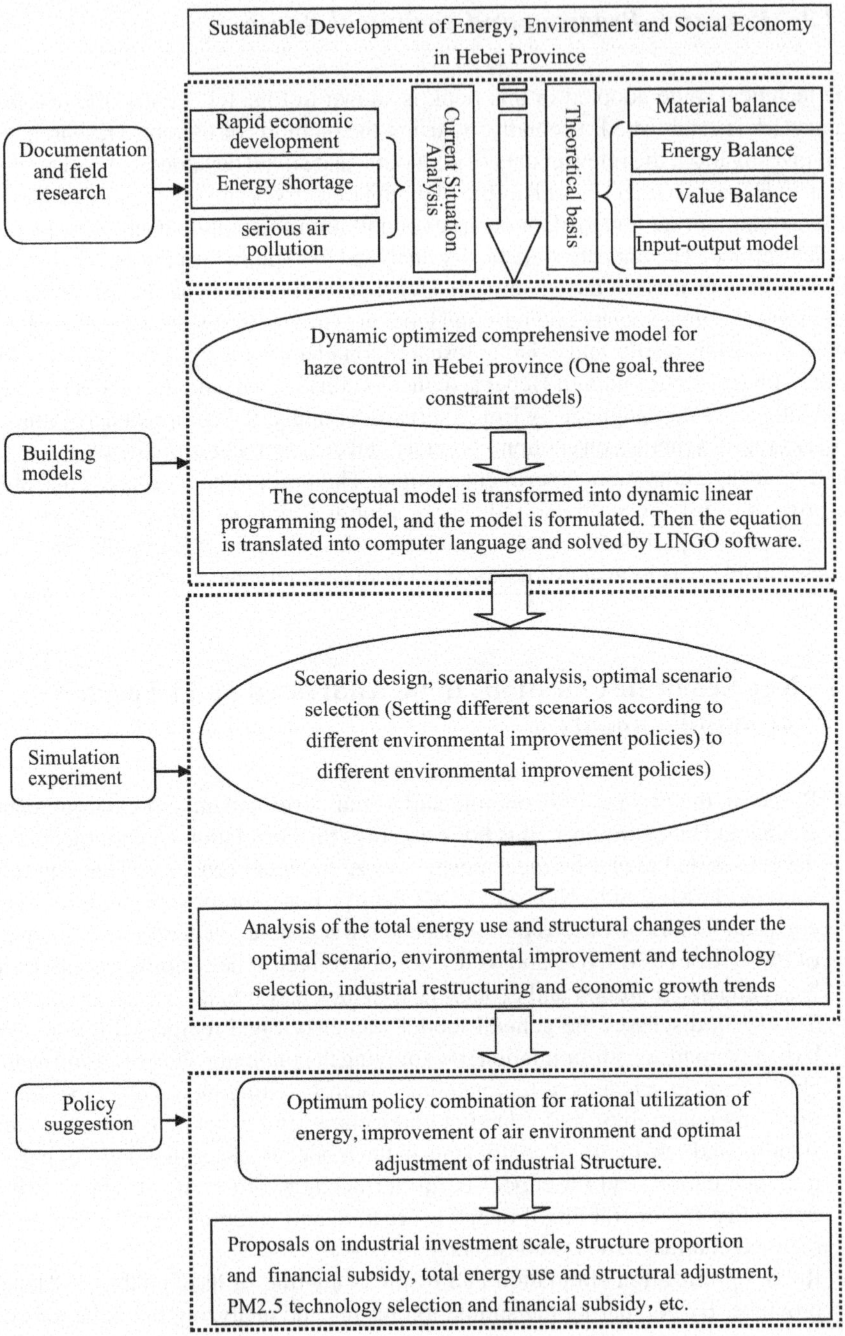

Fig. 1.1 Technology roadmap

implementation scheme of the optimal policy and technology combination at different stages of development is determined.

1.5 Expected Accomplishments

(1) Establish the dynamic optimization model of haze control path.
(2) Propose the best way to control fog and haze in Hebei province and the best implementation plan with both economic and environmental benefits.

Chapter 2
Analysis of the Current Situation of Research Areas

This chapter, which consists of five parts, analyses the current situation of regional economy, society, energy and atmospheric environment, and takes it as the basis of building a comprehensive evaluation model of dynamic optimization of economic, social, energy and atmospheric haze control policies in Hebei province. The first part chooses the administrative divisions and population status of Hebei province to analyze the current social development situation of Hebei province. The second part uses the total economic volume, economic structure and financial revenue and expenditure of Hebei province to analyze the current economic development situation. The third part describes the current energy situation through the energy reserves, total energy utilization, contradiction between energy supply and demand, energy consumption intensity and energy utilization structure of Hebei province. The fourth part analyses the quantity and change of motor vehicles in Hebei province and other cities. The fifth part analyses the present situation, characteristics and countermeasures of haze pollution in Hebei province.

2.1 Administrative Division and Population in Hebei Province

2.1.1 Administrative Division of Hebei Province

Hebei province, with Shijiazhuang as its capital city, is located between 113° 27′ E–119° 50′ E and 36° 05′ N–42° 40′ N, surrounding the capital city, Beijing. The total area of Hebei is 188,800 km^2. Hebei province is 283 km north of Beijing, adjacent to Tianjin in the East and Bohai Sea in the south, Shandong and Henan in the southeast,

G. Zhang et al., *The Effect Evaluation of Haze Governance Policies in Hebei Province–Based on I-O Model*, https://doi.org/10.1007/978-981-15-3757-8_2

Taihang Mountains in the west, Shanxi Province in the northwest, Inner Mongolia Autonomous Region in the north, and Liaoning Province in the northeast.[1]

Hebei province consists of 11 municipal cities (Shijiazhuang, Chengde, Zhangjiakou, Qinhuangdao, Tangshan, Langfang, Baoding, Cangzhou, Hengshui, Xingtai, Handan) and 1 national-level new district (Xiongan New District), 47 municipal districts, 19 county-level cities, 96 counties and 6 autonomous counties (a total of 168 county-level administrative divisions), as shown in Table 2.1.

Hebei province covers an area of 188,800 km^2, accounting for 87% of the total area of BTH region, while Beijing covers an area of 16,400 km^2, accounting for 7.5% of the total area of BTH region, and Tianjin covers an area of 11,900 km^2, accounting for 5.5% of the total area of BTH region.

2.1.2 *Population of Hebei Province*

At the end of 2016, the permanent population of Hebei province was 74.705 million, with an increase of 451,300 million, or 0.61%, than that of 2015. The birth rate was 12.42‰ with 925,000 births, the mortality rate of 6.36‰ with 473,700 deaths. The natural population growth rate was 6.06‰, which is 0.2% higher than the national average level. In terms of gender structure, the male population was 37.96 million, accounting for 50.81% of the total population, while the female population is 36.75 million, accounting for 49.19% of the total population. The gender ratio of the total population was 103.27 (female is 100). From age structure, the population aged 0–17 was 16.2249 million, accounting for 21.72% of the permanent population. The population aged 18–59 was 45.783 million, accounting for 61.29% of the permanent population. The population aged 60 and over was 12.69 million, accounting for 16.99% of the permanent population.

The change of population in Hebei province over the past ten years is shown in Fig. 2.1. Over the past decade, the total population of Hebei province has shown a steady upward trend, with an average growth rate of 0.81%. The average growth rate of rural resident population was −1.91%, and that of urban resident population was 4.01%. The rising proportion of urban resident population was higher than that of rural population, and the process of urbanization is becoming more and more remarkable. In 2016, the rural population was 34.87 million, with a decrease of 1.27 million, or a decrease of 3.51% than that of the same period of previous year. The urban population of Hebei province was 39.83 million, an increase of 1,720,000, or an increase of 4.51% over the same period of last year. Urban population accounted for 53.32% of the total permanent population. The process of population urbanization has been further accelerated.

From the total population of each administrative region, as shown in Fig. 2.2, the population was centralized in the central and southern areas of Hebei province.

[1]The data came from the website of the people's government of Hebei province: http://www.hebei.gov.cn/hebei/10731222/index.html.

Table 2.1 Administrative divisions and areas of cities and municipalities in Hebei province in 2016

Administrative region	Area (km^2)	Administrative region
Shijiazhuang	15,848	Chang′an district, Qiaoxi district, Xinhua district, Jingxing kuang district, Yuhua district, Gaocheng district, Luquan district, Luancheng district, Jingxing county, Zhengding county, Xingtang county, Lingshou county, Gaoyi county, Shenze county, Zanhuang county, Wuji county, Pingshan county, Yuanshi county, Zhao county, Jinzhou city, Xinle city, Xinji county
Tangshan	13,472	Lubei district, Lunan district, Guye district, Kaiping district, Fengnan district, Fengrun district, Caofeidian district, Luanxian county, Luanan county, Leting county, Qianxi county, Yutian county, Zunhua city, Qian′an city
Qinhuangdao	7813	Haigang district, Shanhaiguan district, Beidaihe district, Funing district, Qinglong manchu autonomous county, Changli county, Lulong county
Handan	12,000	Handan mountain district, Congtai district, Renaissance district, Fengfeng mining district, Handan county, Linzhang county, Chengan county, Daming county, Shexian county, Magnetic county, Feixiang county, Yongnian county, Qiu county, Jize county, Guangping county, Guantao county, Quzhou county, Wu′an, Wei county
Xingtai	12,486	Qiaodong district, Qiaodong district, Xingtai county, Lincheng county, Neiqiu county, Baixiang county, Longyao county, Ren county, Nanhe county, Julu county, Xinhe county, Guangzong county, Pingxiang county, Wei county, Qinghe county, Linxi county, Nangong city, Shahe city, Ningjin county
Baoding	22,100	Jingxiu district, Lianchi district, Mancheng district, Qing yuan district, Xushui district, Laishui county, Fuping county, Dingxing county, Tang county, Gaoyang county, Rongcheng county, Lai yuan county, Wangdu county, Anxin county, Yi county, quyang county, Li county, Shunping county, Boye county, Xiong county, Anguo, Gaobeidian, Dingzhou city, zhuozhou
Zhangjiakou	36,860	Qiaodong district, Qiaoxi district, Xuanhua district, Xiayuan district, Xuanhua county, Zhangbei county, Kangbao county, Guyuan county, Shangyi county, Yu county, Yangyuan county, Huaian county, Wanquan county, Zhuolu county, Chicheng county, Chongli county, Huailai county
Chengde	39,519	Shuangqiao district, shuangluan district, yingshouyingzikuang district, chengde county, xinglong county, luanping county, longhua county, fengning manchu autonomous county, kuancheng manchu autonomous county, weichang manchu Mongolian autonomous county, pingquan county

(continued)

Table 2.1 (continued)

Administrative region	Area (km^2)	Administrative region
Cangzhou	13,419	Yunhe district, Xinhua district, Cangxian county, Qingxian county, Dongguang county, Haixing county, Yanshan county, Suning county, Nanpi county, Wuqiao county, Xian county, Mengcun hui autonomous county, Botou, Huanghua, Hejian, Renqiu
Langfang	6500	Guangyang district, Anji district, Gu an county, Yongqing county, Xianghe county, Dasheng county, Wen′an county, Dachang hui autonomous county, Bazhou, Sanhe
Hengshui	8815	Taocheng district, Zaoqiang county, Wuyi county, Wuqiang county, Raoyang county, Anping county, Gucheng county, Fucheng county, Jizhou, Shenzhen, Jingxian county

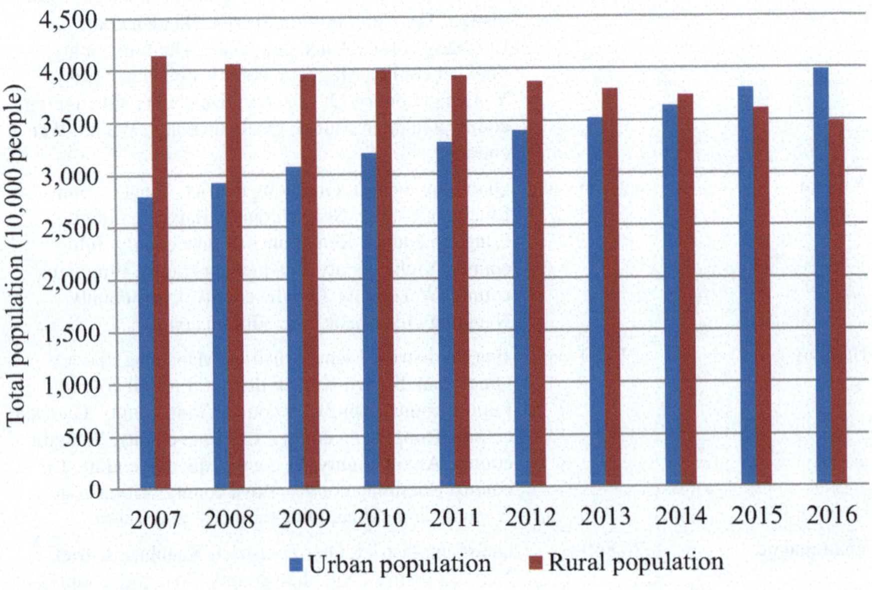

Fig. 2.1 Changes in the total population of urban and rural areas in Hebei province from 2007 to 2016

Baoding City, Shijiazhuang City and Cangzhou City in the central region had a large population, of which Baoding City had 11.6345 million people, accounting for 15.57% of the total population of Hebei province. Shijiazhuang City had 10.784 million people, accounting for 14.44% of the total population of Hebei province. Cangzhou City had 7.5055 million people, accounting for 10.05% of the total population of Hebei province. The most populous areas in the south are Handan City and Xingtai City, of which Handan City had 9.4928 million people, accounting for

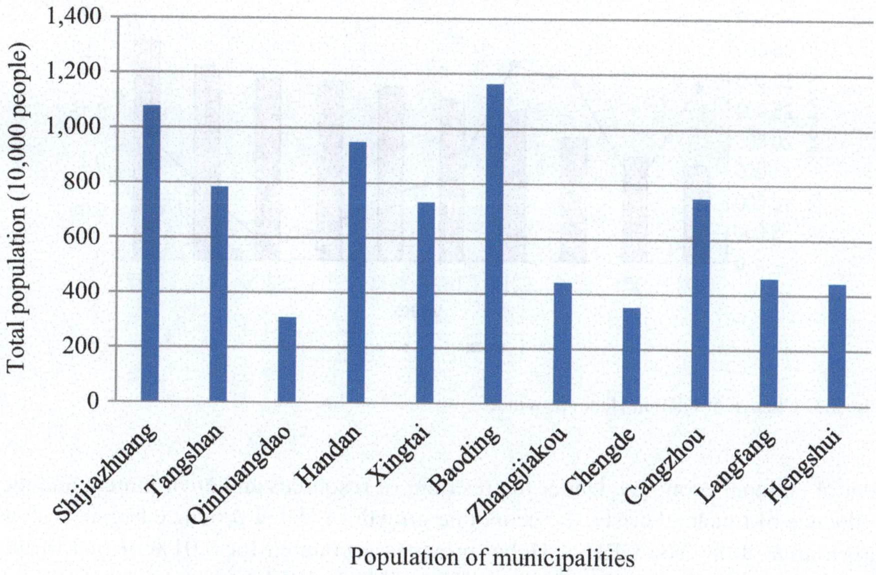

Fig. 2.2 Population of cities in Hebei province in 2016

12.71% of the total population of Hebei province. Xingtai City had 7.3199 million people, accounting for 9.80% of the total population of Hebei province. The most populous area in the eastern region is Tangshan City, which had 7.8436 million people, accounting for 10.50% of the total population of Hebei province. The northern region had a small population, with 4.4251 million people in Zhangjiakou city, accounting for 5.92% of the total population of Hebei province, and 353.18 people in Chengde city, accounting for 4.73% of the total population of Hebei province, while Qinhuangdao city had 3.0946 million people, accounting for 4.14% of the total population of Hebei province.

According to the population distribution of BTH region in 2016, the total population of Hebei province, Beijing, and Tianjin was 74.705 million, 21.729 million, 15.621 million, accounting for 67%, 19%, and 14% of the total resident population of BTH region, respectively.

2.1.3 Analysis of Total Economic Quantity in Hebei Province

The change of GRP of Hebei from 2008 to 2017 is shown in Fig. 2.3. During the Eleventh Five-Year Plan and the Twelfth Five-Year Plan period, the total economic volume of Hebei province increased annually. In 2008, the total economic volume was 1601.2 billion RMB Yuan, and it had increased to 3596.4 billion RMB Yuan by 2017. The average annual growth rate of regional GDP was 9.4, 1.74% lower than

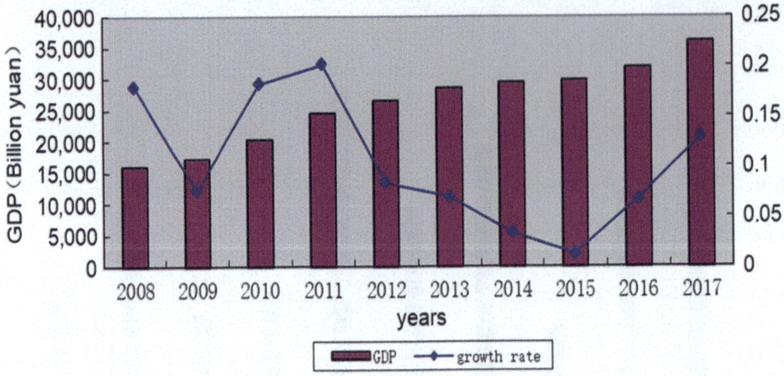

Fig. 2.3 Change of GRP in Hebei province

that of national economy. Under the restraint of resources and environment and the influence of financial crisis, the economic growth of Hebei province began to slow down after 2008. The GRP of Hebei province accounted for 5.01% of that of the whole country in 2008, while it dropped to 4.35% in 2017.

Although the economy of Hebei province is growing annually, the gaps between its cities are simultaneously expanding, and the problem of uncoordinated regional economic development is becoming increasingly prominent. The economic development of Shijiazhuang and Tangshan is relatively faster. In 2017, Tangshan's GRP was 710.61 billion RMB Yuan, accounting for 19.8% of Hebei's GRP. Shijiazhuang's GRP was 600.35 billion RMB Yuan, accounting for 16.7% of Hebei's GRP. By contrast, the economic development of Qinhuangdao and Hengshui is slow. In 2017, Qinhuangdao's GRP was 150.6 billion RMB Yuan, accounting for 4.2% of the GRP of Hebei province. The GRP of Hengshui City was 155.01 billion RMB Yuan, accounting for 4.3% of the GRP of Hebei province. As shown in Table 2.2.

According to the economic development of Beijing, Tianjin and Hebei in 2017.

The GRP of Hebei, Beijing, Tianjin was 3596.4 billion RMB Yuan, 2800.04 billion RMB Yuan, and 1858.53 billion RMB Yuan, accounting for 44%, 34%, and 22% of that of BTH region, respectively.

2.1.4 Economic Structure of Hebei Province

With the rapid economic development, the industrial structure of Hebei province is also constantly optimized. The added value of primary industry in 2007 was 180.47 billion RMB Yuan, which had increased to 350.79 billion RMB Yuan by 2017, with the industrial share decreasing from 13.26% in 2007 to 9.75% in 2017. The added value of the secondary industry in 2007 was 720.20 billion RMB Yuan, which had increased to 1741.65 billion Yuan by 2017, with the industrial share decreasing from 52.93% in 2007 to 48.43% in 2017. The added value of the tertiary industry in 2007

Table 2.2 GRP of Hebei province in 2017

Municipal cities	GRP (100 million RMB Yuan)	Per capita GRP (RMB Yuan)	Proportion (%)
Tangshan	7106.1	78525.27	19.8
Shijiazhuang	6003.5	51248.09	16.7
Cangzhou	3816.9	43940.34	10.6
Handan	3666.3	33554.87	10.2
Baoding	3227.3	26112.04	9.0
Langfang	2880.6	53118.23	8.0
Xingtai	2236.36	24254.79	6.2
Zhangjiakou	1555.6	30843.04	4.3
Chengde	1618.6	38517.80	4.5
Qinghuangdao	1506	40804.05	4.2
Hengshui	1550.1	27580.59	4.3

was 460.07 billion RMB Yuan, which had increased to 1503.96 billion RMB Yuan by 2017, and the proportion of industry increased from 33.81% in 2007 to 41.82% in 2017 (Figs. 2.4 and 2.5).

1. **Primary industry**

Hebei is an important grain and cotton producing region in China. The main grain crops are wheat, corn, sorghum, millet and potatoes. Cotton is the main cash crop, and Hebei province is an indispensable cotton production base in China. In addition, oilseeds, hemps, sugar beets, and tobacco are also important, and together with cotton are the five largest cash crops in the province. Animal husbandry is the second most

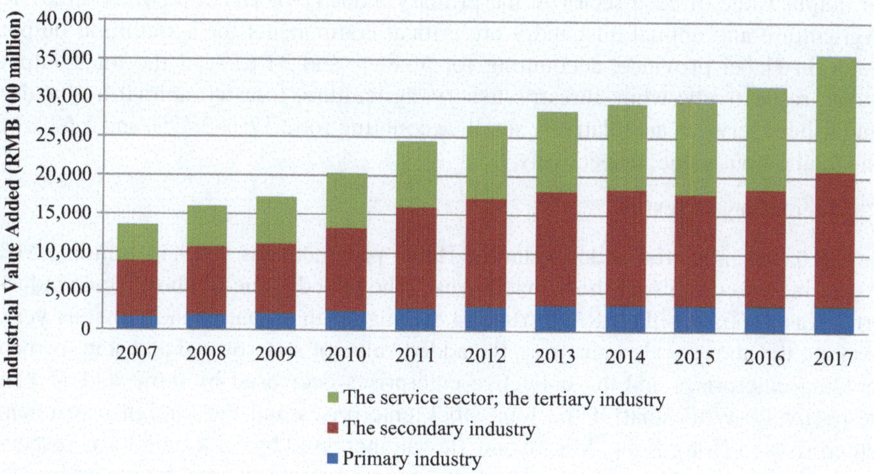

Fig. 2.4 Added value of three industries in Hebei province from 2007 to 2017

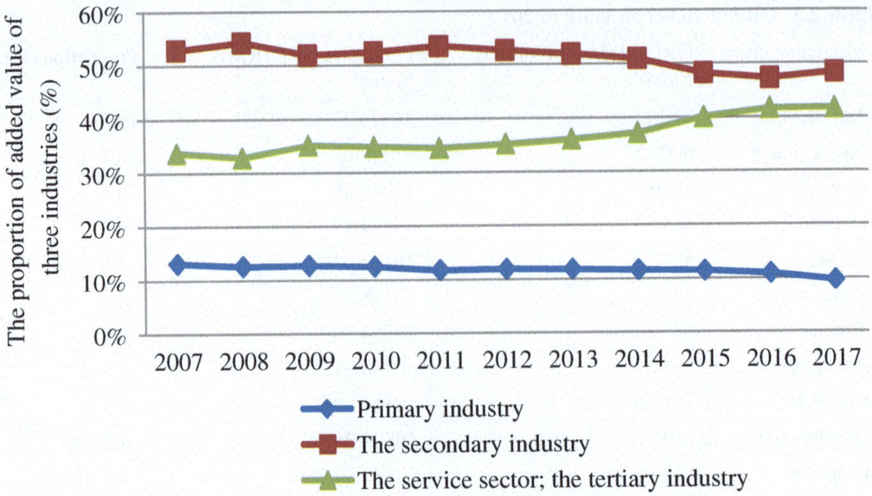

Fig. 2.5 Changes of industrial structure in Hebei province from 2007 to 2017

important agricultural sector in Hebei province after farming. In 2016, the sown area of grain in Hebei province was 6.327 million hectares, 1% lower than that of the previous year, accounting for more than 80% of the total cultivated land. The total grain output rose by 2.9% to 34.60 million tons. The cotton planting area was 289,000 hectares, 19.7% lower than that of previous year, and the total output was 300,000 tons, 19.8% lower than that of 2015. Oil seeding area increased by 1.5% to 468,000 hectares, and the total production increased by 3.3% to 1.565 million tons, while vegetable seeding area decreased by 0.5% to 1.236 million hectares, and its total output decreased by 0.6% to 81.934 million tons. Figure 2.6 shows the proportion of output value of each sector in the primary industry of Hebei province in 2016. Agriculture and animal husbandry are critical components for agricultural output value in Hebei province, accounting for 56.86% and 31.87% of the total output value, respectively, while forestry, fishery, agriculture, forestry, animal husbandry and fishery services are relatively small, accounting for 2.17%, 3.47% and 5.62% of the total output value, respectively.

2. **Secondary industry**

In 2016, the industrial added value of Hebei province was 1319.44 billion RMB Yuan, increased 4.6% of the previous year. The added value of above scale industries was 1166.38 billion RMB Yuan, 4.8% higher than that of the previous year. Among the above scale industries, the added value of state-owned and state-owned holding enterprises and the collective enterprises decreased by 0.6% and 15.8%, respectively. While that of the joint-stock enterprises and the foreign investment enterprises in Hong Kong, Macao and Taiwan increased by 5.7% and 3.4%, respectively. In addition, the added value of the iron and steel industry decreased by 2% over the previous year. Fortunately, the added value of the equipment manufacturing

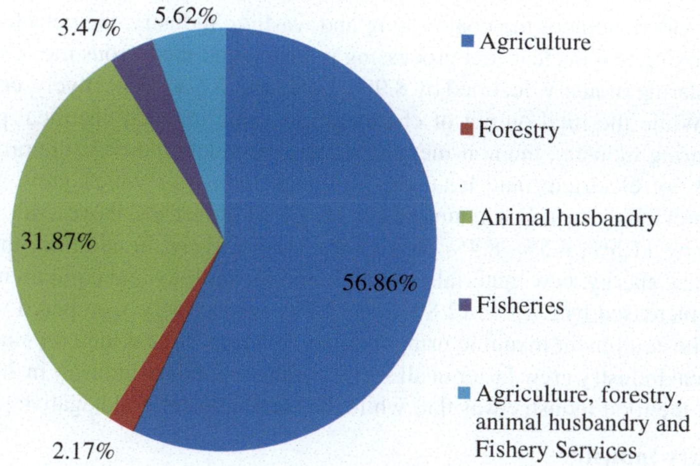

Fig. 2.6 Value-added ratio of primary industry sectors in Hebei province in 2016

industry, the petrochemical industry, the pharmaceutical industry, the construction material industry, food industry, and the textile and garment industry increased by 10.02%, 4.9%, 5.4%, 3.2%, 3.2%, and 6.3% than that of the previous year, respectively (see Fig. 2.7). The added value of the six high energy-consuming industries increased by 1.4%, while the growth rate fell by 1.8% over the previous year. Among

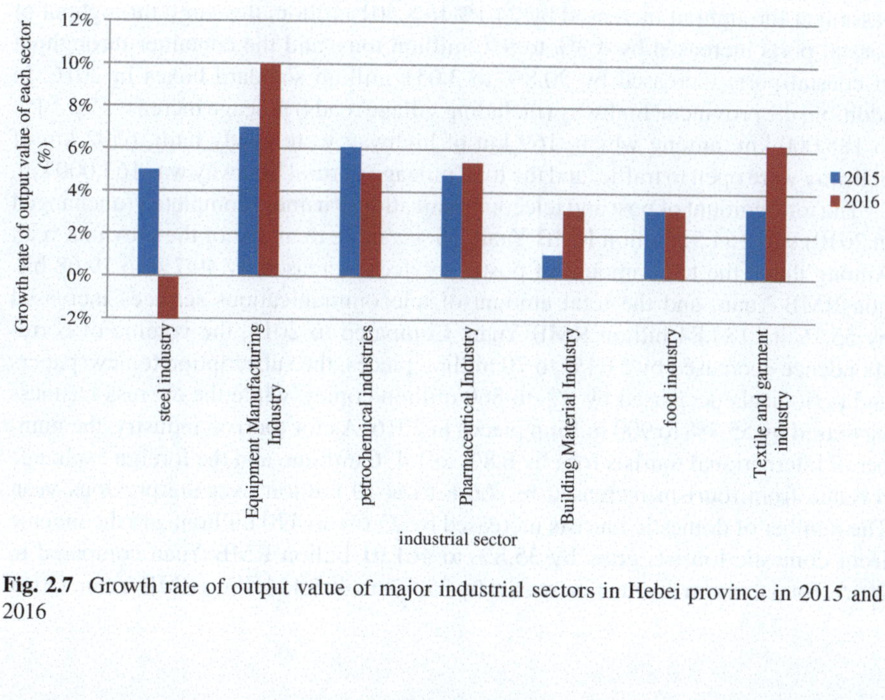

Fig. 2.7 Growth rate of output value of major industrial sectors in Hebei province in 2015 and 2016

them, the added value of the coal mining and washing industry, the petroleum processing, coking and nuclear fuel processing industry, and the ferrous metal smelting and calendaring industry declined by 8.9%, 3.6%, and 2.5%, respectively, compared to 2015. While the total output of chemical raw materials and chemical products manufacturing industry, the non-metallic mineral products industry, the production and supply of electricity and heat, the high-tech industries which grew by 2.4% over the previous year and accounting for 18.4% of industries above scale in 2016 increased by 11.5%, 4.5%, 8.4%, and 13.0%, respectively. In addition, the added value of new energy, new materials and high-end technology and equipment manufacturing increased by 27.5%, 12.8% and 12.7%, respectively. Compared with that in 2015, the equipment manufacturing industry, pharmaceutical industry and textile and garment industry grew faster in all sectors of the secondary industry in 2016. The growth of the food industry was flat, while the steel industry had negative growth.

3. **Tertiary Industry**

The tertiary industry of Hebei province has developed rapidly in recent years, but the proportion of the tertiary industry is still relatively low compared with that of Beijing and Tianjin. In 2016, the output value of tertiary industry was 1332.07 billion RMB Yuan, accounting for 41.4% of the GRP of Hebei. Compared with 2015, the added value of transportation, warehousing and postal industry increased by 0.4% to 236.93 billion RMB Yuan, the total volume of cargo transportation increased by 6.0% to 2.11 billion tons, the turnover of cargo increased by 2.7% to 1233.93 billion TKM, the total volume of passenger transport dropped by 4.6% to 510 million people, the turnover of passenger transport dropped by 2.1% to 123.81 billion kilometers, the airport passenger throughput increased by 24.1% to 8.501 million, the cargo throughput of coastal ports increased by 4.3% to 950 million tons, and the container throughput of coastal ports increased by 20.8% to 3.051 million standard boxes in 2016. In addition, the provincial highway (including village roads) mileage increased by 2.1% to 188,000 km, among which, 169 km of highway were newly built, 6502 km of highway were open to traffic, and the total mileage of rural highway was 163,000 km.

The total amount of post and telecommunications business completed (unchanged in 2010) was 154.52 billion RMB Yuan, 55.4% more than that of the previous year. Among them, the total amount of postal services increased by 49.7% to 19.68 billion RMB Yuan, and the total amount of telecommunications services increased by 55.7% to 134.84 billion RMB Yuan. Compared to 2015, the volume of correspondence decreased by 51.1% to 70 million pieces, the subscription to newspapers and periodicals decreased by 8% to 860 million copies, while the express business increased by 55.7% to 900 million pieces in 2016. As for tourism industry, the number of international tourists rose by 6.8% to 1.476 million, and the foreign exchange revenue from tourism increased by 7.6% to $670 million over the previous year. The number of domestic tourists increased by 25.6% to 470 million, and the income from domestic tourists grew by 35.8% to 461.01 billion RMB Yuan compared to 2015. Total tourism revenue increased by 35.6% to 465.45 billion RMB Yuan. Other sectors of tertiary industry, such as science, education, culture and health, have also developed rapidly.

From the perspective of industrial structure development in BTH region, the proportion of three industries in 2017 in Hebei province, Beijing, and Tianjin was 9.8:48.4:41.8, 0.4:19:80.6, and 1.2:40.8:58, respectively.

2.1.5 *Fascial Revenue and Expenditure of Hebei Province*

Fascial revenue in Hebei province has steadily increased. The total fiscal revenue of Hebei province from 2012 to 2017 was 347.926 billion RMB Yuan, 365.24 billion RMB Yuan, 376.456 billion RMB Yuan, 406.511 billion RMB Yuan, 437.34 billion RMB Yuan and 508.69 billion RMB Yuan, respectively. The annual growth rates were 15.30%, 4.98%, 3.07%, 7.98%, 7.58% and −26% respectively, with an average annual growth rate of 7.86%.

The general budget expenditure of Hebei province increased steadily, too. From 2012 to 2017, the total fiscal expenditure of Hebei province was 407.944 billion RMB Yuan, 440.958 billion RMB Yuan, 467.73 billion RMB Yuan, 563.219 billion RMB Yuan, 603.8 billion RMB Yuan and 661.52 billion RMB Yuan, respectively. And the annual growth rates were 15.32%, 8.09%, 6.07%, 20.41%, 7.21% and 9.56% respectively, with an average annual growth rate of 10.15%.

In 2017, under the complex and severe situation, with the correct leadership of provincial Party committees and provincial government, departments at all levels in Hebei province earnestly implemented the spirit of General Secretary Xi Jinping's series of important speeches, adhered to the general tone of work of striving for progress steadily, focused on improving the quality and efficiency of economic development, actively adapted to the new normal of economic development, sought actively and steadily increase, adjusted the structure, promoted reform, controlled pollution, benefited the people, prevented risks, strived for stability and progress in the national economy, pursued innovation in stability, made progress in stability, and were determined to make comprehensive progress in social undertakings. During the 12th Five-Year Plan period, Hebei province increased infrastructure construction, expanded financial expenditure, and promoted the harmonious development of economy and society. Under the situation of extremely sharp contradictions between revenue and expenditure, people's livelihood security was given priority, and a large number of facts of benefiting the people were effectively implemented. Figures 2.8 and 2.9 show the changes of fiscal revenue and expenditure in Hebei province from 2012 to 2017.

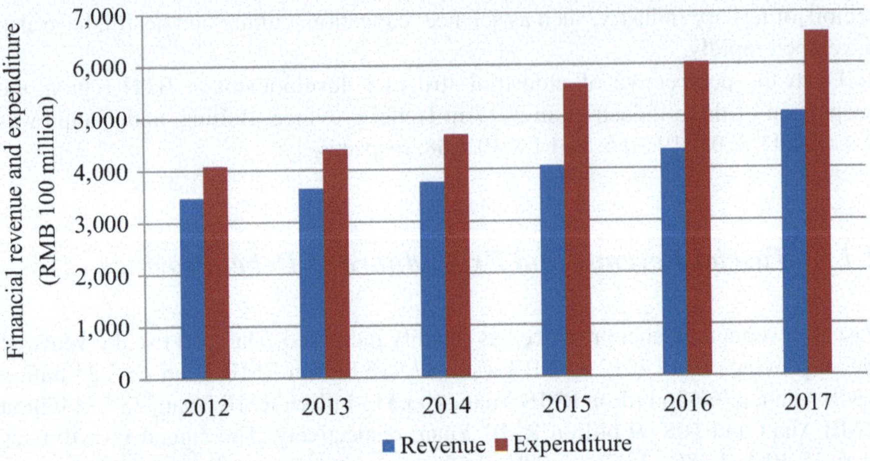

Fig. 2.8 Financial revenue and expenditure of Hebei province from 2012 to 2017

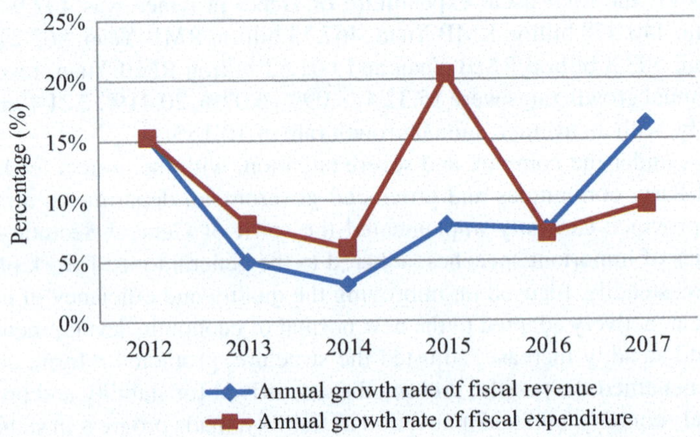

Fig. 2.9 Annual growth rate of fiscal revenue and expenditure in Hebei province from 2012 to 2017

2.2 Analysis of Energy Situation in Hebei Province

2.2.1 Analysis of Energy Gross Product

Hebei province, known as "Yanzhao Coal Warehouse", is one of the main energy supply bases in China, and it is also one of the areas with the earlies development of modern energy industry in China. By the end of 1996, coal had accumulated proven reserves of 16.72 billion tons and retained reserves of 14.728 billion tons, ranking 11th in China. Over 80% of the total reserves in the province are distributed

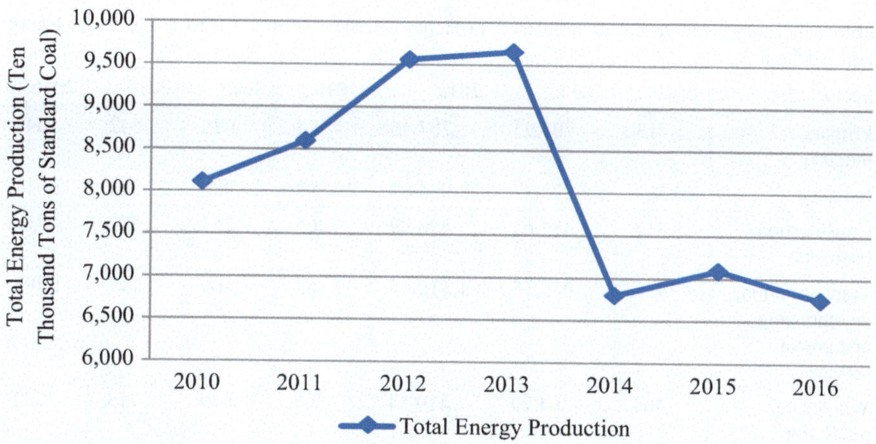

Fig. 2.10 Change of energy production in Hebei province from 2010 to 2016

in Tangshan, Handan, Xingtai and other places. Oil and natural gas resources are also relatively abundant, which are stored in Jizhong, Dagang and Jidong oilfields. By the end of 1998, the total proved oil reserves was 10.88 tons, which ranked 5th in the country. The cumulative proven reserves of natural gas are 17.8 billion cubic meters. Abundant energy reserves play an indispensable role in stabilizing energy supply and ensuring economic development in Hebei province. The total energy production of Hebei province in 2010–2016 is shown in Fig. 2.10. The impact of environmental carrying capacity on the total energy supply in Hebei province decreased annually since 2013. In 2013, the total energy production in Hebei province was 96.5642 million tons of standard coal, and by 2016 it had reduced by 30% to 67.53 million tons of standard coal.

2.2.2 *Analysis of Total Energy Utilization*

With the rapid development of Hebei's economy, the total energy utilization increased by 13.71% from 262.014 million tons of standard coal in 2012 to 297.94 million tons of standard coal in 2016. The proportion of energy consumption in industry was relatively high. In 2010, 205.63 million tons of standard coal was consumed by industry, accounting for 78.5% of the total consumption. While that in 2016 was 220.14 million tons of standard coal, accounting for 73.89% of the total consumption. The proportion of energy consumption in other industries is relatively small. From 2010 to 2016, the energy utilization of various industries in Hebei province is shown in Table 2.3.

Table 2.3 Energy utilization of industries in Hebei province from 2010 to 2016. Unit: 10,000 Standard Coal

Sector	2010	2011	2012	2013	2014	2015	2016
Primary industry	713	703.93	707.46	574	625	642	648
Industry	20,563	23,275.37	24,086.65	23,389	22,785	22,184	22,014
Construction business	319	384.62	396.99	265	253	297	312
Transportation, warehousing and postal service	971	1,075.15	1,118.65	1,162	1,109	1,111	1,286
Wholesale, retail and accommodation, catering	304	364.79	419.54	565	639	713	772
Other	716	785.63	846.24	829	911	1,057	1,134
Life consumption	2,615	2,499.49	2,674.61	2,881	2,997	3,391	3,628
Total energy consumption	26,201	29,498.29	30,250.21	29,664	29,320	29,395	29,794

According to the total energy utilization in BTH region in 2016, the total energy utilization in Hebei province, Beijing, and Tianjin was 297.944 million tons of standard coal, 63.595 million tons of standard coal, and 82.445 million tons of standard coal, accounting for 67%, 14% and 19% of the total energy utilization in BTH region, respectively.

2.2.3 *Analysis of the Contradiction of Energy Supply and Demand*

Although Hebei province produces a lot of energy, compared with the increasing energy consumption, it cannot meet its own economic and social development needs. In 2010, the total energy consumption of Hebei province was 262.0 million tons of standard coal, which was 3.23 times of the total energy production, and 69% of the energy resources had to be transferred from outside. In 2016, the total energy consumption of Hebei province was 292.9 million tons of standard coal, which was 4.33 times of the total energy production, and 77% of the energy were transferred from outside. The gap between energy supply and demand was further widened and the contradiction between supply and demand was prominent (see Fig. 2.11).

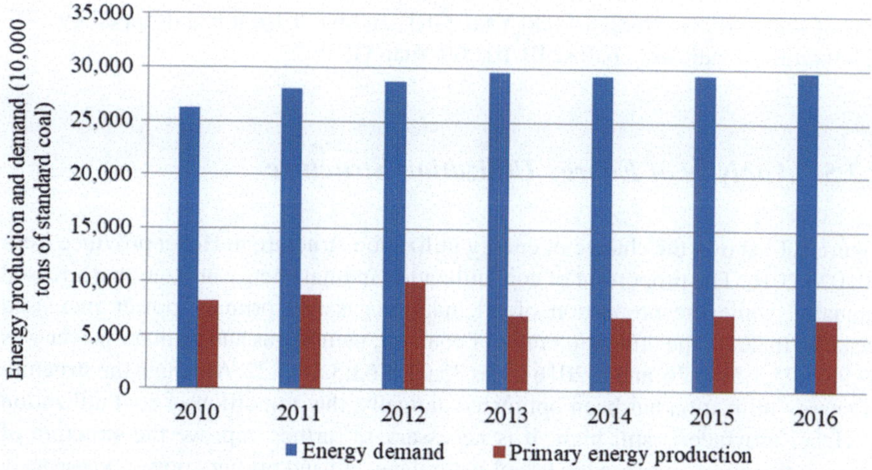

Fig. 2.11 Energy supply and demand in Hebei province from 2010 to 2016

2.2.4 *Analysis of Energy Consumption Intensity*

In recent years, Hebei province has intensified the adjustment of industrial structure and energy structure, scientifically planned energy utilization and introduced advanced energy-saving technologies, and greatly improved energy efficiency. Figure 2.12 shows the change of energy consumption intensity in Hebei province from 2010 to 2016. From 2010 to 2016, the intensity of energy consumption in Hebei province decreased annually. In 2010, the intensity of energy consumption was 1.28

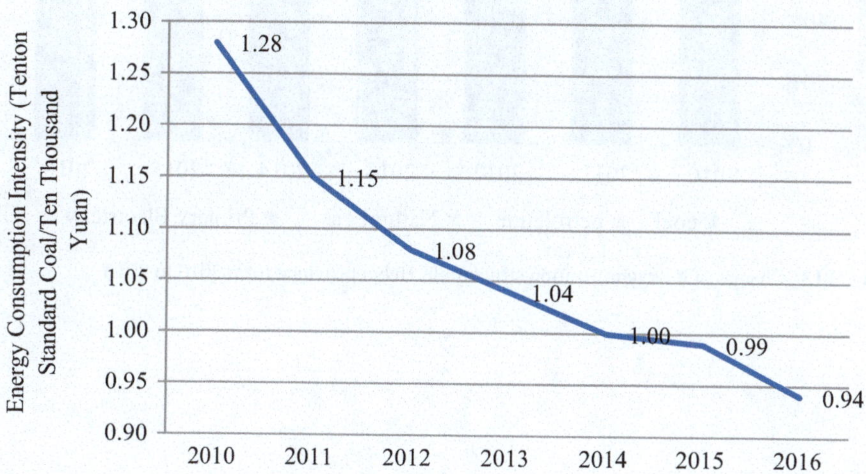

Fig. 2.12 Energy consumption intensity of Hebei province from 2010 to 2016

tons of standard coal/RMB 10,000 Yuan GRP, and by 2016, it had dropped by 27% to 0.94 tons of standard coal/RMB 10,000 Yuan GRP.

2.2.5 *Analysis of Energy Utilization Structure*

Figure 2.13 shows the change of energy utilization structure in Hebei province from 2010 to 2016. The proportion of coal utilization in total energy utilization decreased annually, while the proportion of oil, natural gas and primary power increased steadily. In 2010, the utilization ratio of coal, oil, natural gas and primary power was 89.98:7.75:1.51:0.76, and in 2016 it was 85.01:8.63:3.14:3.22. Although the structure of energy utilization has been optimized annually, the proportion of coal utilization in Hebei province is still high. It is necessary to further improve the structure of energy utilization, increase the use of natural gas, oil and primary power, vigorously develop clean energy and increase the proportion of clean energy utilization.

According to the energy utilization structure of BTH region, in 2016, the proportion of coal, oil, natural gas, primary power and other energy resources utilization in Hebei province was 89.98:7.75:1.51:0.76, while that in Beijing was 9.81:32.93:31.68:25.58, and that in Tianjin city was 61.96:26.71:0.9:10.43.

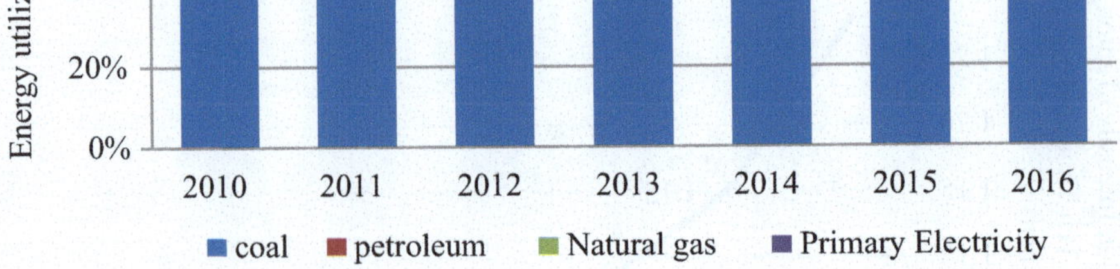

Fig. 2.13 Change of energy utilization structure in Hebei province from 2010 to 2016

2.3 Current Situation of Motor Vehicle Development in Hebei Province

2.3.1 *Motor Vehicle Ownership in Hebei Province*

Automobile industry is an essential pillar industry of our national economy. It plays an important role in stimulating economic growth, increasing employment, increasing revenue and taxation. By 2016, Hebei province had 18.666 million motor vehicles, 2.0711 million operating motor vehicles, 7.5842 million cars, 7.3138 million private cars, 12.917 million cars, 1.6992 million motorcycles, 0.298 million highway buses, 0.0026 million school buses, and 0.0263 million new energy vehicles which include 25,000 pure electric vehicles and 130,000 hybrid electric vehicles.

In recent years, the total number of vehicles in Hebei province is on the rise. The total number of motor vehicles increased at an average annual growth rate of 5.75%, and additional 5.3218 million motor vehicles were supplied in 2016, compared with 2010. In 2014, the total number of motor vehicles decreased by 3.76% compared to 2013. However, the growth rate of the total number of motor vehicles increased significantly after 2014, as shown in Fig. 2.14.

Figure 2.15 shows the proportion of various types of vehicles in Hebei province in 2016, with passenger cars accounting for 36.44%, motorcycles 25.74%, trucks for 9.84%, tractors for 10.75%, trailers for 2.07% and other types of vehicles accounting for 15.16%.

From the perspective of motor vehicles ownership in BTH region in 2016, the number of motor vehicles in Hebei, Beijing and Tianjin was 18.666 million, 5.72 million, and 2.74 million, accounting for 69, 21, and 10% of the total motor vehicles ownership in BTH region.

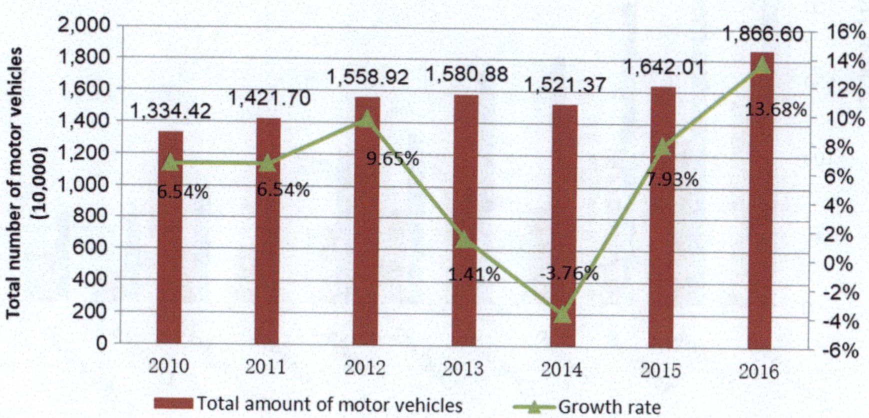

Fig. 2.14 Total motor vehicle volume and growth rate in Hebei province from 2010 to 2016

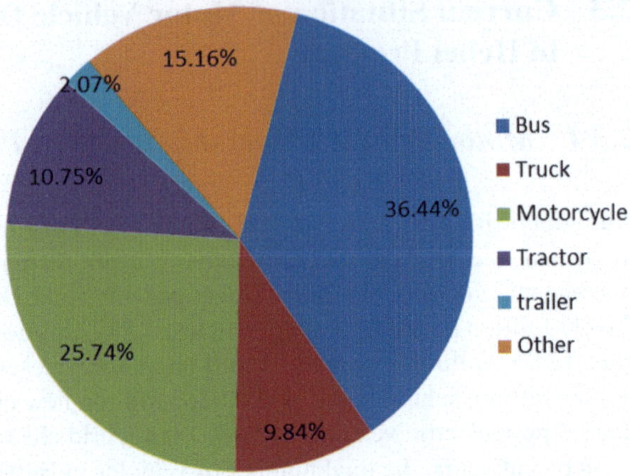

Fig. 2.15 Proportion of various motor vehicles in Hebei province in 2016

2.3.2 Vehicle Ownership in Hebei Province

The regional difference of motor vehicle ownership in Hebei province is large. Figure 2.16 shows the number of motor vehicles, cars and private cars in Hebei province in 2016. Shijiazhuang, Baoding and Tangshan had relatively high motor vehicle ownership, while Hengshui, Zhangjiakou, Chengde and Qinhuangdao had

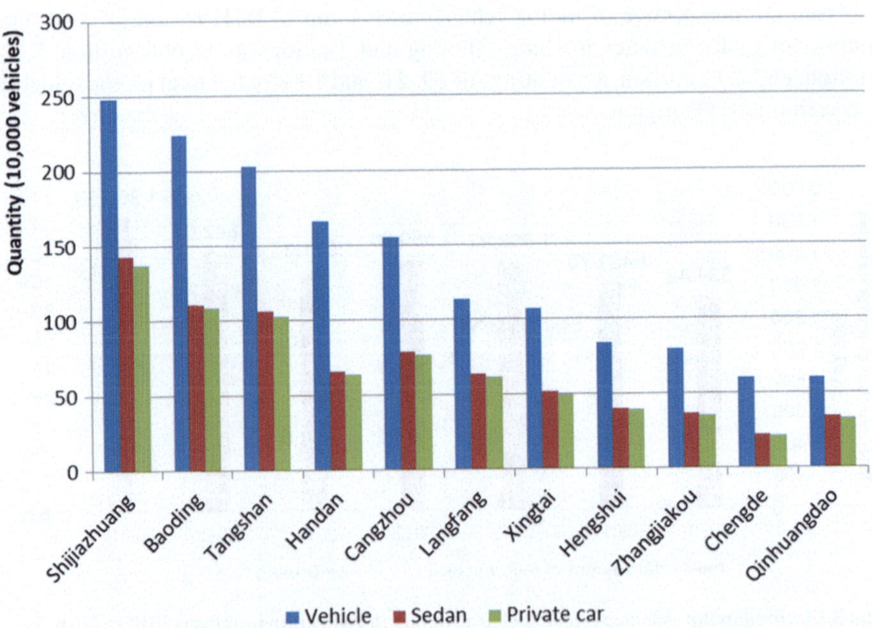

Fig. 2.16 The number of motor vehicles, cars and private cars in Hebei province in 2016

relatively small proportion of motor vehicle ownership, and there are great gaps in the total amount of motor vehicles between the cities in Hebei.

2.4 Current Situation of Haze Pollution in Hebei Province and Countermeasure

2.4.1 Haze Pollution in Hebei Province

With the rapid development of China's economy and the increasing population, environmental problems have become increasingly prominent. According to the national air quality data of 2016 published by the Ministry of Environmental Protection in February 2017, of the 74 cities monitored nationwide for air quality in 2016, six of the 10 cities with the worst air quality belonged to Hebei. From 2013 to 2016, the average annual air quality standard rate of Hebei province was 46%. The increasing environmental problems seriously affected people's health and life happiness.

Air quality has improved slightly, but the overall situation is still not optimistic. From 2013 to 2016, the fine days of grade II air quality in districts and cities of Hebei province were 129 days, 152 days, 190 days and 207 days, respectively. The average days of severe pollution and above were 80 days, 66 days, 36 days and 33 days, respectively. With the joint efforts of the government and people of Hebei province, the number of fine days increased annually, and the number of severe pollution days decreased annually (see Fig. 2.17), but the number of non-excellent days still accounted for 40% of the whole year.

Although the concentration of pollutants has decreased, it is still high. In 2013, there were seven cities including Shijiazhuang, Tangshan, Langfang, Baoding,

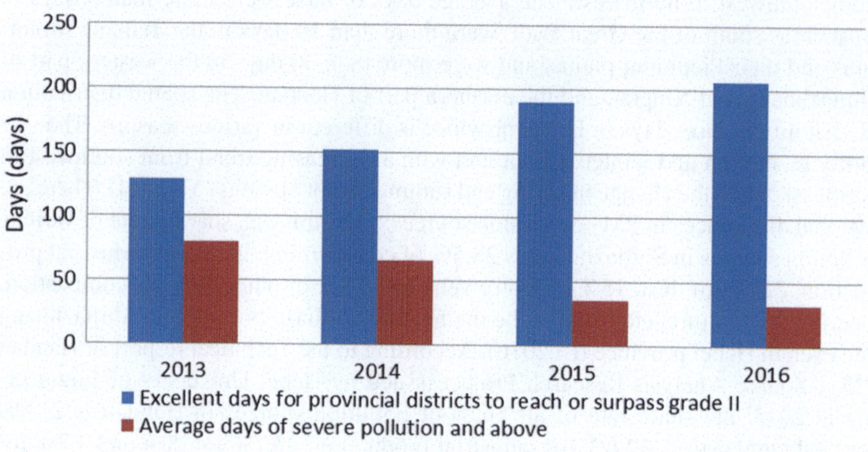

Fig. 2.17 Days of achievement and heavy pollution in Hebei province from 2013 to 2016

Table 2.4 $PM_{2.5}$ average values of cities in Hebei province in the first half of 2013–2016

$PM_{2.5}$-year mean (μg/m^3)	2013	2014	2015	First half of 2016
Shijiazhuang	148.5	122.6	87.2	72.9
Chengde	51.5	53.5	41.9	37.5
Zhangjiakou	43.1	34.3	34.3	28.5
Qinhuangdao	65.2	59	47.6	43
Tangshan	114.2	98.4	84.7	66.8
Langfang	113.8	99.3	85	57.6
Baoding	127.9	127.2	107	83.2
Cangzhou	93.6	88	70.3	59.8
Hengshui	120.6	107.6	99.2	85.1
Xingtai	155.2	131.4	101.7	79
Handan	127.8	114.2	91.6	71.1

Hengshui, Xingtai and Handan in Hebei province had average annual $PM_{2.5}$ emission exceeding 100. In 2014, the environmental situation of Hebei province improved slightly. The average $PM_{2.5}$ of Tangshan and Langfang decreased to less than 100. In 2015, the average $PM_{2.5}$ of Hebei province reached more than 100. By the first half of 2016, every city in Hebei had average annual $PM_{2.5}$ emission below 100, as shown in Table 2.4.

2.4.2 *Characteristics of Haze Pollution in Hebei Province*

The average days of fog and haze in Hebei province showed a decreasing trend from southwest to northwest. The average days of haze were more than 5 days in most areas south of the Great Wall, were more than 10 days in the Taihang Mountains and their Piedmont plains, and were more than 30 days in the western part of Shijiazhuang and Xingtai, and the northern part of Handan. The spatial distribution of visibility in haze days in Hebei province is different in various seasons. The visibility in autumn and winter is poor and with a decreasing trend from southwest to northeast, while the change in spring and summer is not obvious (Yu 2014). There are regional differences in $PM_{2.5}$ emission sources. The emission sharing rate of various pollution sources in Shijiazhuang is 28.5% of coal burning, 25.2% of industrial production, 22.5% of dust, 15% of motor vehicles, 8.8% of other biomass combustion, catering, agriculture, etc. Coal is the main source of haze pollution in Shijiazhuang and even in Hebei province (Li 2016). According to the Technical Report of Handan $PM_{2.5}$ Source Analysis Research Project issued by Hebei University of Engineering in 2015, the share rate of all kinds of pollution sources in Handan is 21.8% for coal combustion, 32.6% for industrial production, 23.7% for dust and 4.7% for motor vehicles. Industrial production is the main source of haze pollution in Handan. "Chengde Atmospheric $PM_{2.5}$ Source Analysis Research Report" published in 2016 by Chengde Environmental Protection Bureau and Beijing University of Technology shows that the main source of $PM_{2.5}$ emission in Chengde is coal combustion (11.2%), industrial production (28.3%), dust (23.1%), motor vehicles (9.8%), biomass combustion, construction material industry catering and other (7%), with industrial production as the main source.

2.4.3 Measures for Haze Pollution Control in Hebei Province

Faced with the serious environmental problems in Hebei province, the State and Hebei Provincial Government have issued a lot of policies and regulations to control environmental pollution, protect and improve the environment, and protect public health and safety, which including water and soil conservation policy, industrial protection policy, new energy automobile subsidy policy, emission trading policy, energy saving and emission reduction policy, introduction of pollutant treatment technology policy, regulation policies for enterprises with high pollution, high energy consumption and high emissions, restrictions on motor vehicles, clean energy utilization and other environmental protection policies. The model in this book selected and simulated the policies of soil and water conservation motor vehicle restriction, introduction of $PM_{2.5}$ treatment technology, industrial structure adjustment, new energy vehicle subsidy and development and utilization of clean energy. The specific content of countermeasures for fog and haze control in Hebei province are as follows.

The administrative measures for the prevention and control of radioactive pollution in Hebei province contain 27 articles, which were adopted at the 48th executive meeting of the Hebei provincial government on October 26, 2001, and are have been promulgated and implemented since then. The Measures are formulated for the prevention and control of radioactive pollution, the protection and improvement of the environment, and the protection of public health and safety.

Regulations on the administration of environmental protection of construction projects in Hebei province with a total of six chapters and 36 articles, were adopted at the 24th Meeting of the Standing Committee of the Eighth People's Congress of Hebei province on December 17, 1996, revised at the 10th meeting of the Standing Committee of the Tenth People's Congress of Hebei province on July 22, 2004, and amended at the 13th meeting of the standing committee of the 10th people's congress of Hebei province on January 9, 2005. In order to strengthen the environmental protection management of construction projects in Hebei province and prevent environmental pollution and ecological damage of construction projects, the Regulations are formulated in accordance with laws and regulations such as the Environmental Protection Law of the People's Republic of China and the actual situation of Hebei province.

The regulations of Hebei province on environmental protection, which consist of six chapters and 51 articles, were amended and adopted at the 14th session of the standing committee of the tenth people's congress of Hebei province on March 25, 2005, and came into force as of May 1, 2005. In order to protect and improve the living and ecological environment, prevent and control pollution and other environmental hazards, the Regulations are formulated in accordance with the provisions of the Environmental Protection Law of the People's Republic of China, the Water Pollution Prevention Law of the People's Republic of China and the Air Pollution Prevention Law of the People's Republic of China, and other laws and administrative regulations, in combination with the actual situation of the province.

The administrative measures for environmental protection industries of Hebei province, which contains 19 articles, is amended in accordance with Order No. 5 of the provincial government (2007) and came into force on April 22, 2007 in accordance with serial No. 79 of the notice of general office of Hebei people's government on promulgation of catalogue of regulations of the current effective provincial government issued on November 15, 2007 by general office of Hebei provincial government. In order to strengthen the supervision and management of environmental protection industry, promote the healthy development of environmental protection industry and protect and improve the environment, the measures are formulated in accordance with the Regulations on Environmental Protection of Hebei province and the relevant provisions of the State and in light of the actual situation of the province.

On May 31, 2010, the Circular on the Pilot Subsidies for Private Purchase of New Energy Vehicles was jointly issued by the Ministry of Finance, the Ministry of Science and Technology, the Ministry of Industry and Information Technology and the National Development and Reform Commission. In order to implement the relevant decisions of the 91st Standing Meeting of the State Council, accelerate the technological progress of the automobile industry, focus on fostering strategic emerging industries, and promote energy conservation and emission reduction, the Ministry of Finance, the Ministry of Science and Technology, the Ministry of Industry and Information Technology and the National Development and Reform Commission carried out pilot work on subsidies for private purchases of new energy vehicles according to the "request for instructions on expanding energy-saving and new-energy vehicle demonstration promotion" approved by the state council (Caijian [2010] No. 41). Based on the automotive industry foundation, residents' purchasing power and other relevant requirements, the four ministries and commissions selected five cities to prepare the pilot implementation plan of private purchase of new energy vehicles subsidy, and organized experts to demonstrate the implementation plan, and then started the pilot after the demonstration.

The Notice of the Measures for the Management of Transaction of Emission Rights of Major Pollutants in Hebei province (for trial implementation) with a total of 28 articles was issued by the Hebei Provincial Government on December 28, 2010, and came into effect on May 1, 2011. On the premise of meeting the requirements of environmental quality and total emission control of major pollutants, the transferor or transferee of the right to discharge major pollutants may, at the trading institution, openly trade the annual permitted discharge of major pollutants obtained according to law, and the annual allowable emissions of major pollutants for new, renovated and expanded projects that need to be approved after May 1, 2011 must be obtained through trading. The measures are formulated in accordance with the relevant provisions of the State in order to accelerate the transformation of the mode of economic development, promote the reduction of major pollutants, implement and standardize the trading activities of major pollutant emission rights.

The 12th Five-Year Plan for Energy Conservation and Emission Reduction was issued by the State Council on August 6, 2012. The Plan called on all parts of the country to effectively reduce the total amount of major pollutants. The plan clearly pointed out that it would promote the prevention and control of motor vehicle pollution, and promote the development of strategic emerging industries such as new energy vehicles. By 2015, the average fuel consumption of passenger cars would be reduced from 8 L/100 km in 2010 to 6.9 L/100 km.

The "ten measures for the prevention and control of air pollution" was finalized at an executive meeting of the state council on June 14, 2013. The measures include reducing pollutant emissions, strictly controlling new energy consumption in high-energy and high-pollution industries, vigorously promoting cleaner production, speeding up the adjustment of energy structure, strengthening the constraints of energy-saving and environmental protection indicators, implementing a new energy-saving and emission-reduction mechanism with incentives and constraints, intensifying the collection of pollutant discharge fees, and increasing credit support for prevention and control of air pollution.

The implementation plan of air pollution control in Shijiazhuang (2013–2017) was formulated and issued by the Shijiazhuang government on June 18, 2013. The plan clearly stated that after five years of efforts, the city's air quality would be significantly improved and the number of days with heavily polluted weather would be greatly reduced. The government would complete the relocation and transformation of polluting enterprises, formulate a new relocation plan for urban polluting enterprises, continue to give preferential policies for the relocation of polluting enterprises during the "three-year great change" period, encourage and guide polluting enterprises to speed up the relocation and transformation.

The Circular on the Issuance of the Action Plan for the Prevention and Control of Atmospheric Pollution was issued by the state council (Guo fa [2013] No. 37) on September 10, 2013. The circular contains ten parts: Strengthening comprehensive control and reducing emissions of multiple pollutants; Adjusting and optimizing the industrial structure to promote industrial transformation and upgrading; Speeding up the technological transformation of enterprises and improving the ability of scientific and technological innovation; Accelerating the adjustment of the energy mix and increase the supply of clean energy; Strict energy conservation and environmental protection access and optimizing the industrial space layout; Promoting market mechanisms to play its role and improving environmental and economic policies; Improving the system of laws and regulations, and exercising strict supervision and management in accordance with the law; Establishing a monitoring, early warning and emergency response system to cope with heavy pollution; Establishing a monitoring, early warning and emergency response system, properly responding to heavy pollution, clarifying the responsibilities of government, enterprises and society; Mobilizing the whole people to participate in environmental protection.

The Implementation Plan of Hebei Air Pollution Prevention and Control Action Plan was issued by Hebei Provincial Party Committee and Hebei Provincial People's Government on September 12, 2013. Guided by the scientific concept of development, the Plan vigorously promoted the construction of ecological civilization, regarded good ecological environment as an important goal of improving people's livelihood, and adhered to economic development and environmental protection, comprehensive promotion and key breakthroughs, consistency between territorial management and regional cooperation, synchronization of total emission reduction and quality improvement, transformation of development mode, optimization of industrial structure, efforts to solve air pollution problems with fine particulate matter ($PM_{2.5}$) as the focus, highlighting the pollution control of key cities, key industries and key enterprises, the formation of government leadership, enterprise governance, the formation of a new mechanism for air pollution prevention and control led by the government, administered by enterprises, driven by innovation, supervised by the society and participated by the public, the overall goal of the province's environmental air quality to improve significantly by 2017, making an important contribution to the capital and the surrounding air quality improvement.

The Rules for the Implementation of the Action Plan for the Prevention and Control of Atmospheric Pollution in Beijing, Tianjin, Hebei and Surrounding Areas were issued on 17 September 2013 by the Ministry of Environmental Protection, the National Development and Reform Commission, the Ministry of Industry and Information Technology, the Ministry of Finance, the Ministry of Housing and Urban-Rural Construction and the State Energy Administration. The Rules focused on improving regional environmental air quality, reducing heavy pollution weather, taking multiple measures to strengthen the prevention, the control of air pollution in winter, and comprehensively reducing regional pollution discharge load.

The Regulations on Public Participation in Environmental Protection of Hebei province with a total of six chapters and 43 articles were adopted at the 11th meeting of the Standing Committee of the 12th People's Congress of Hebei province on November 28, 2014. In order to protect the public's right to know, participate and supervise environmental protection, promote the construction of ecological civilization and create a good production and living environment, the Regulations were formulated in accordance with the relevant laws and administrative regulations such as the Environmental Protection Law of the People's Republic of China and the actual situation of the province.

The Regulations on the Prevention and Control of Environmental Pollution by Solid Waste in Hebei province with a total of six chapters and 46 articles were adopted at the 14th meeting of the Standing Committee of the 12th People's Congress of Hebei province on March 26, 2015, and came into effect on June 1, 2015. Relevant provisions are made on the prevention and control of environmental pollution by solid waste, supervision and management of environmental pollution by solid waste, prevention and control of environmental pollution by hazardous waste, and legal liability. In order to prevent and control solid waste pollution to the environment, protect public health, maintain ecological safety and promote sustainable economic and social development, the Regulations are based on the provisions of the Environmental Protection Law of the People's Republic of China and the Environmental Prevention and Control Law of the People's Republic of China on Solid Waste Pollution, and the actual formulation of the province.

"Beijing-Tianjin-Hebei Regional Environmental Protection Pioneer Breakthrough Cooperation Framework Agreement" (hereinafter referred to as the Agreement) was officially signed on December 2, 2015 by the environmental protection departments of Beijing, Tianjin and Hebei. The Agreement clearly focused on the prevention and control of air, water and soil pollution, and took joint legislation, unified planning, unified standards, unified monitoring and coordinated pollution control as breakthroughs to jointly prevent and control pollution and jointly improved the regional ecological environment quality. The signing of the Agreement means that the three places have taken another substantive step in implementing the spirit of the Beijing-Tianjin-Hebei Cooperative Development Plan, speeding up the breakthrough in the field of ecological environmental protection, and jointly building the Beijing-Tianjin-Hebei Demonstration Zone for Ecological Restoration and Environmental Improvement.

The Regulations on the Prevention and Control of Air Pollution in Hebei province with a total of eight chapters and 93 articles were adopted by the Fourth Session of the Twelfth People's Congress of Hebei province on January 13, 2016, and came into effect on March 1, 2016. According to the idea of "source control, process control and consequence punishment", the work of atmospheric pollution prevention and control was comprehensively standardized, and the responsibilities of the government and its departments, prevention and control measures of various pollution sources, coping mechanism of heavy polluted weather and regional cooperation mechanism were stipulated. In order to prevent and control air pollution, protect and improve living environment and ecological environment, safeguard human health, and promote the development of socialist modernization, the Regulations are formulated in accordance with relevant laws and regulations such as the Law of the People's Republic of China on the Prevention and Control of Air Pollution, and in light of the actual situation of the province.

The Circular on Organizing and Declaration of Investment Options in the Central Budget of Pollution Control Projects in Beijing, Tianjin, Hebei and Key Areas was issued by the General Office of the State Development and Reform Commission on April 12, 2016. The Circular includes a pilot project of comprehensive prevention and control of atmospheric haze and third-party treatment of environmental pollution. It is limited to the centralized treatment of industrial wastewater and industrial point source treatment projects determined in the pilot implementation scheme of third-party environmental pollution control approved by Chengde City, Hengshui New Industrial Zone, Handan New Chemical Park and Xinji Leather City Leather Industry Park.

On June 14, 2016, the Shijiazhuang Municipal People's Government promulgated the Work Program for Air Pollution Prevention and Control 2016, with the purpose of resolutely reducing the total amount of coal burning, strictly controlling dust pollution, strengthening the control of motor vehicle pollution, vigorously promoting enterprise emission reduction, accelerating enterprise relocation and vigorously carrying out tree planting and greening, strengthen emergency response to heavy pollution weather and improve environmental supervision in an all-round way.

The Regulations on Environmental Protection and Governance in Rural Areas of Hebei province with a total of seven chapters and 45 articles were adopted by the Standing Committee of the Twelfth People's Congress of Hebei province at its 22nd Meeting on July 29, 2016, and came into effect on October 1, 2016. In order to protect and improve the rural environment, promote the construction of livable villages, advocate a civilized and healthy way of production and life, and promote sustainable economic and social development, the Regulations are formulated in accordance with the relevant laws and administrative regulations such as the Environmental Protection Law of the People's Republic of China and the actual situation of the province.

On November 10th, 2016, the Hebei Provincial Government promulgated the Dispatching Order to carry out the environmental protection duties of the relevant departments responsible for the supervision and management of air pollution prevention and control. It mainly includes 7 parts as follows: (1) For cities that have lagged behind in fulfilling the annual target task of air pollution prevention and control, or those are expected to be unable to meet the targets for improving the quality of the air environment, the key tasks such as the reduction of relevant industrial enterprises in the area, the control of urban dust, the restriction of motor vehicles, and the prevention and control of non-point source pollution are scheduled according to the results of pollution source analysis and judgment. (2) During the heating period and under unfavorable weather conditions, the effect of emergency response is not remarkable, the AQI (air quality index) remains high, and the enterprises of high-pollution and high-emission industries in the dispatching area effectively implement the measures of reducing atmospheric pollutants. (3) When regional heavy pollution occurs and has obvious regional transmission characteristics, targeted dispatching and transmission channel cities (Shijiazhuang, Tangshan, Langfang, Baoding, Cangzhou, Hengshui, Xingtai, Handan, Dingzhou and Xinji) jointly implement atmospheric pollutant reduction measures. (4) During the periods when major events are held in Beijing, Tianjin, Hebei and surrounding cities, jointly implement measures should be taken for reducing atmospheric pollutants in order to ensure the air quality of the hosting sites, the surrounding areas of Beijing, Tianjin and Hebei. (5) The problem of environmental pollution is prominent, and the effect of local remediation is not obvious. Consequently, the industries which have a greater impact on the quality of the atmospheric environment should be scheduled. (6) The prominent environmental problems should be founded and handled by the State and Provincial Party committees and provincial governments in their supervision and inspection, as well as the environmental problems with strong public reaction, concentrated reporting and protracted decision-making. (7) Other important matters concerning the necessary dispatch of air pollution prevention and control should be dispatched after research.

"Eighteen Strengthening Measures for Dust Prevention and Control in Construction of Hebei province" (known as the "Strong Eighteen") was issued by Hebei Housing and Construction Department on December 20, 2016. The "Strong Eighteen" clearly stated that all construction enterprises must set up public signs for dust prevention and control in obvious locations at the entrances and exits of the construction sites, including the names of construction, construction, supervision and supervision enterprises, the names of the responsible persons for dust prevention and control, contact telephone and reporting telephone.

On March 31, 2017, the Opinions on Promoting Comprehensive Control of Air Pollution was put forward by the Hebei Provincial Party Committee and Provincial Government. The Opinion adhered to the people-centered development thinking, adhered to the two bottom lines of ecology and development, focused on accelerating the improvement of air quality, implemented scientific haze management, coordinated haze management, and iron-handed haze management, and resolutely fought for the blue sky.

On November 17, 2016, Shijiazhuang People's Government urgently promulgated the "Implementation Plan for the Action of Clean Swords" and planned to take 45 days to carry out the action of air pollution prevention and control of Clean Swords in Shijiazhuang. According to the Plan, from November 17 to December 31, 2016, the main urban area would be restricted by single and double number motor vehicles, during which urban buses would be free to ride. After being submitted to the provincial government for approval, all administrative departments and institutions in the city implemented the "nine to five" system of staggered commuting.

The 13th Five-Year Plan for Eco-environmental Protection of Hebei province was issued by the Hebei Provincial Government on March 7, 2017. According to the plan, by the end of 2017, the task of "6643" to eliminate excess capacity would be completed, and cleaner production audits in cement, chemical, petrochemical and non-ferrous industries would be basically completed. The coal production capacity of the whole province would be controlled to 70 million tons by 2020, and the production capacity of steel, cement and flat glass would be controlled to 200 million tons, 200 million tons and 200 million weight boxes, respectively.

On April 1, 2017, the Hebei Provincial Government formulated the Special Implementation Plan for Coal Reduction in the Thermal Power Industry. Hebei would implement the project of eliminating backward production capacity in the thermal power industry, shut down all 20 enterprises and 181.4 million kilowatt units whose operation period is less than 300,000 kW, 15 enterprises and 135,000 kW units with pure condensation units under 300,000 kilowatt level. For small enterprises that can be covered by large heating enterprises, the shutdown would be carried out in an orderly manner after implementing alternative heating schemes orderly. At the same time, Hebei had proposed that all coal-fired thermoelectric units in urban areas would be replaced by clean fuels such as natural gas. For urban coal-fired thermal power units that do not have clean fuel substitution, the relocation would be carried out steadily on the basis of properly resolving personnel placement, asset disposal, site selection of alternative projects and supporting pipeline network.

On August 18, 2017, the Ministry of Environmental Protection, the Development and Reform Commission and other 16 departments formulated the Action Plan for the Comprehensive Control of Air Pollution in autumn and winter of 2017–2018 in Beijing, Tianjin, Hebei and the surrounding areas. The Action Plan adhered to the problem-oriented approach and took the results of the comprehensive control of "scattered and dirty" enterprises and clusters and the stable discharge of elevated sources as the firm positions, focused on the reduction of coal pressure reduction, standard upgrading, peak shift production, and took proper response to severely polluted weather as an important breakthrough. It strengthened joint prevention, control, strict law enforcement supervision, and accountability of inspectors. In addition, the plan comprehensively implemented tough operations, and mobilized the entire population to jointly respond to severely polluted weather.

On May 8, 2018, Langfang Environmental Protection Bureau formulated the "Implementation Plan of Air Pollution Prevention and Control in Langfang City 2018". The plan adhered to the combination of engineering treatment and fine management and control, the combination of key areas and key period management and control, and the combination of territorial responsibility and department management. It is clear that the annual average concentration of $PM_{2.5}$ in Langfang City had reached below 58 micrograms per cubic meter, and the city's air quality had consistently and steadily exited the "Top 10" in the 74 air quality rankings of the country and was better than the 2017 annual target.

On July 19, 2018, the General Office of the People's Government of Shijiazhuang issued the Action Plan for the Special Rehabilitation of Disorganized and Polluted Enterprises in Shijiazhuang City, which requires counties (cities and districts) to carry out a new round of "Disorganized and Polluted" Enterprise Investigation on the basis of consolidating the effect of last year's rectification. The plan had clarified the definition criteria of "scattered and polluted" enterprises, continued to carry out the reinvestigation of "scattered and polluted" enterprises, in accordance with the three ways of "shutting down and banning, upgrading and transforming, integration and relocation", to achieve "resolutely ban, rectification standards, relocation has a place to go", and effectively eliminate potential rebound.

The Technical Guidelines for Environmental Impact Assessment (HJ2.2-2018) were implemented on December 1, 2018, and issued by the Ministry of Ecological Environment of the People's Republic of China on July 30, 2018. The technical guidelines provided guidance for implementing the "Environmental Protection Law of the People's Republic of China" and the "Environmental Impact Assessment Law of the People's Republic of China", preventing air pollution, promoting air quality improvement, and further standardizing the atmospheric environmental impact assessment of construction projects.

"Shijiazhuang City Environmental Air Dust Monitoring Work Implementation Plan" was released on August 3, 2018 by Shijiazhuang Environmental Protection Bureau. This program monitored 23 counties in Shijiazhuang, and selected the dust reduction monitoring points based on the networked township monitoring stations. It is a critical basis for scientifically and accurately reflecting the amount of dust reduction in the city, and an important support for the precise implementation of dust pollution prevention and control.

The publication time, department and policy name of each policy are shown in Table 2.5.

Table 2.5 Haze control policies in Hebei province from 2001–2018

Department	Time of promulgation or implementation	Policy name
Hebei Provincial Government	26 October, 2001	Measures for the Prevention and Control of Radioactive Pollution in Hebei province
Hebei Provincial Government	9 January, 2005	Regulations on Environmental Protection Management of Construction Projects in Hebei province
Hebei Provincial Government	25 March, 2005	Regulations on Environmental Protection of Hebei province
Hebei Provincial Government Office	22 April, 2007	Measures for the Management of Environmental Protection Industry in Hebei province
Ministry of Finance, Ministry of Science and Technology, Ministry of Industry and Information Technology, National Development and Reform Commission	31 May, 2010	Notice on Pilot Subsidies for Private Purchase of New Energy Vehicles
Hebei Provincial Government	28 December, 2010	Measures for the Management of Transaction of Emission Rights of Major Pollutants in Hebei province
The State Council	6 August, 2012	Twelfth Five-Year Plan for Energy Conservation and Emission Reduction
The State Council	14 June, 2013	Ten Measures for Prevention and Control of Air Pollution
Shijiazhuang Government	18 June, 2013	Implementation Plan of Air Pollution Control in Shijiazhuang City
The State Council	10 September, 2013	Notice on the Issuance of an Action Plan for the Prevention and Control of Air Pollution
Hebei Provincial Government	12 September, 2013	Implementation Plan of the Action Plan for the Prevention and Control of Air Pollution in Hebei province

(continued)

Table 2.5 (continued)

Department	Time of promulgation or implementation	Policy name
Ministry of Environmental Protection, National Development and Reform Commission, Ministry of Industry and Information Technology, Ministry of Finance, Ministry of Housing and Urban-Rural Construction, State Energy Administration	17 September, 2013	Rules for the Implementation of the Action Plan for the Prevention and Control of Atmospheric Pollution in Beijing, Tianjin, Hebei and the Surrounding Areas
Hebei Provincial Government	28 November, 2014	Regulations on Public Participation in Environmental Protection of Hebei province
Hebei Provincial Government	26 March, 2015	Regulations on the Prevention and Control of Environmental Pollution by Solid Wastes in Hebei province
Environmental Protection Departments in Beijing, Tianjin and Hebei	2 December, 2015	Beijing-Tianjin-Hebei Regional Environmental Protection First Breakthrough Cooperation Framework Agreement
Hebei Provincial Government	13 January, 2016	Regulations on the Prevention and Control of Air Pollution in Hebei province
Office of National Development and Reform Commission	12 April, 2016	Notice on Organizing the Declaration of Investment Options in the Central Budget of 2016 for Pollution Control Projects in Beijing, Tianjin, Hebei and Key Areas
Shijiazhuang Municipal Government	14 June, 2016	Shijiazhuang Air Pollution Prevention and Control Action Program 2016
Hebei Provincial Government	29 July, 2016	Regulations on Environmental Protection and Governance in Rural Areas of Hebei province
Hebei Provincial Government	7 November, 2016	Dispatching Order
Shijiazhuang People's Government	17 November, 2016	On the Implementation Plan of the Action of Cutting Off Defamation with Swords

(continued)

Table 2.5 (continued)

Department	Time of promulgation or implementation	Policy name
Hebei Provincial Housing and Construction Department	20 December, 2016	18 Strengthening Measures for Dust Control in Construction of Hebei province
Hebei Provincial Government	7 March, 2017	13th Five-Year Plan for Eco-environmental Protection in Hebei province
Hebei Provincial Government	31 March, 2017	Opinions on Strongly Promoting Comprehensive Control of Air Pollution
Hebei Provincial Government	1 April, 2017	Special Implementation Plan for Coal Reduction in Hebei Thermal Power Industry
Ministry of Environmental Protection	18 August, 2017	Action Plan for Comprehensive Control of Air Pollution in Autumn and Winter of 2017–2018 in Beijing, Tianjin, Hebei and Its Surrounding Areas
Langfang Environmental Protection Bureau	8 May, 2018	Implementation Plan of Air Pollution Prevention and Control in Langfang City in 2018
General Office of Shijiazhuang Municipal People's Government	19 July, 2018	Shijiazhuang City 2018 Disorganized and Polluted "Enterprises Special Renovation Action Plan"
Ministry of Ecological Environment of the People's Republic of China	30 July, 2018	Technical Guidelines for Environmental Impact Assessment Atmospheric Environment (HJ2.2-2018)
Ministry of Ecological Environment of the People's Republic of China	3 August, 2018	Shijiazhuang City Environmental Air Dust Monitoring Work Implementation Plan

2.5 Summary

In recent years, the economy of Hebei province has grown rapidly, and the process of industrialization and urbanization has been further accelerated. Increasing population and rapid economic growth put tremendous pressure on energy and environmental carrying capacity. The unreasonable structure of industry and energy utilization, and the rapid growth of motor vehicle ownership have become the main reasons for the formation of fog and haze in Hebei province.

In order to improve the air quality of the environment and promote the high-quality economic development of Hebei province, a series of policy documents has been issued, but the comprehensive impact of these policies on the economy, society and environment of Hebei province remain undiscovered

Hebei province accounts for 87% of the area of BTH region, 67% of the permanent population, 44% of the total economy, 67% of the total energy use and 69% of the motor vehicle ownership. From the perspective of comprehensive management of fog and haze in BTH region, the management of fog and haze in Hebei province plays an essential role in the comprehensive management of fog and haze in BTH region.

Chapter 3
Dynamic and Optimal Comprehensive Evaluation Model of Haze Control Policy in Hebei Province

Based on the analysis of the current situation of economic and social development, energy utilization, vehicle ownership, haze pollution and control in Hebei province, this chapter constructs a dynamic optimization model of environmental and economic impact of haze control policy in Hebei province by using input-output table, and based on the theory of material balance, value balance and energy balance and multi-objective optimization method. The comprehensive evaluation model can scientifically reflect the material flow, value flow, energy flow and their coupling relationship generated by economic and social activities in Hebei province. Computer simulation (Lingo) is used to calculate the optimal solution, and predict the impact of haze control policies on economic, social and environmental of Hebei province in various scenarios. This chapter mainly includes three parts: theoretical basis, conceptual model and basic assumptions, and model building.

3.1 Theoretical Basis of Model Construction

3.1.1 Theory of Material Balance, Value Balance and Energy Balance

Material balance is an important concept in environmental economics. According to the theory of material balance, the process of production and consumption of economy is a series of physical and chemical reactions, which follows the law of conservation of mass. There is a material flow relationship between the environmental system and the economic system. Without accumulation, the material input from the environmental system to the economic system would be greater than that from the economic system to the environmental system. The material balance model can be expressed by the equation $ES = EI + K$, in which ES is the material input of the environmental system to the economic system, EI is the pollutant discharged from the economic system to the environmental system, and K is the material accumulation

G. Zhang et al., *The Effect Evaluation of Haze Governance Policies in Hebei Province–Based on I-O Model*, https://doi.org/10.1007/978-981-15-3757-8_3

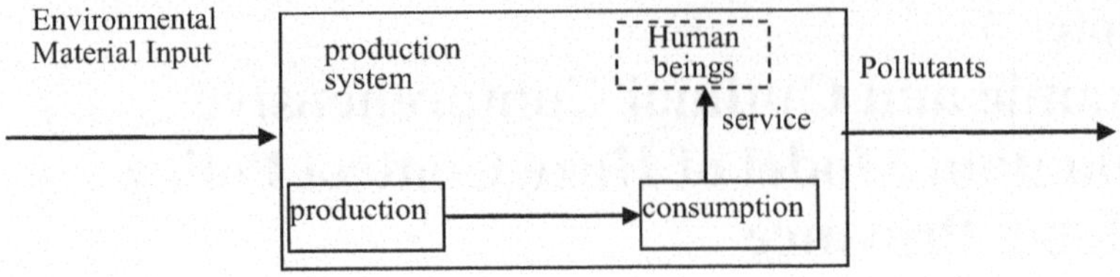

Fig. 3.1 Relationship between material flow in environmental system and economic system

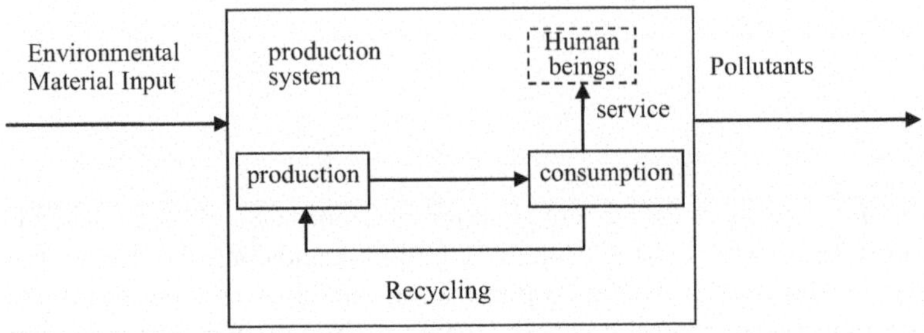

Fig. 3.2 Material flow in environmental and economic systems with recycling

of the economic system. If there is no accumulation in the process of economic production and consumption, i.e. $K = 0$, the input environmental substances will inevitably return to the environment in the form of pollutants (Fig. 3.1). The only function of the environment in this process of material flow is to provide services for human beings, as shown in the dotted line section of Fig. 3.1. If there is material accumulation in real economic production and consumption activities, which means $K > 0$, there will be recycling. Material becomes a part of raw materials and is reused (Fig. 3.2).

Energy balance is the relationship between energy supply and energy demand in the process of economic and social production and consumption. In the analysis of energy supply and demand, various forms of energy are converted into unified energy units based on the law of energy conservation. The most common energy balance analysis tool in modern economy is the energy balance table, which is composed of individual balance tables of various kinds of energy. It is in the form of matrix tables. The data of various kinds of energy resources supply, processing conversion and terminal consumption are summarized and recorded in several tables, and the reporting period is described intuitively. The relationship between the supply and demand of various kinds of energy and their processing and transformation, as well as the structure of resource supply and consumption demand in the whole country or region are shown in a whole table. Using the energy balance table, we can intuitively reveal the balance relationship between energy resources, conversion and terminal consumption in quantity. Quantitative and systematic analysis of regional energy flow process using energy balance can scientifically describe the internal relationship mechanism between energy, economy, society and ecological environment in a certain space-time range, and help to put forward corresponding policy and technical control countermeasures.

Value balance is the balance of the total value, composition and change of the total social product among various departments. Value balance can be reflected by the value input-output table. Using value-based input-output model to analyze the value balance not only unifies the unit of measurement of the model, but also reflects all economic activities of material and non-material, tangible and intangible, physical and virtual in social production, and unifies the material flow and value flow in economic activities, so as to reveal the whole economy more intuitively.

3.1.2 Multi-objective Optimization Model

Multi-objective optimization model is widely used in industry, agriculture, transportation, commerce, national defense, construction, communications, government departments and other fields of practical work. It mainly solves the problem of finding the maximum and minimum of the function, such as the optimal production plan, the optimal allocation, the optimal design, the optimal decision-making and the optimal management. Optimizing problems involve a variety of contents and are very complex, but they all have the same key factors: variables, constraints and objective functions.

In the optimization problem, the function of the extremum (or maximum and minimum) to be solved is called the objective function. Common equation of objective function is shown as:

$$Maximize/Minimize = f(x) = f(x_1, x_2 \ldots x_m)$$

Variables refer to the uncertain quantities related to constraints and objective functions in the optimization problem. The variables involved in the problem are $x_1, x_2 \ldots x_m$. When the extremum of objective function is obtained, the constraints that variables must satisfy are called constraints. Generally speaking, there are two kinds of constraints in mathematical language description:

That is, equality constraints:

$$g_i(x) = 0\, i = 1, 2, \ldots m$$

Inequality constraints:

$$h_i(x) \geq 0\, i = 1, 2, \ldots m$$

Or. $h_i(x) \leq 0\, i = 1, 2, \ldots m$.

The multi-objective optimization model mainly solves two problems. The first is to find the minimum or extreme value of the function under certain constraints; the second is to find the value of variables when the extreme value is obtained.

In establishing the comprehensive evaluation model of environmental economy, we can not only set economic growth as the objective function and environmental pollution level as the constraint condition, discuss the optimal development of economy under the existing carrying capacity of resources and environment, but also set the minimum environmental pollution as the objective function to study the optimal development under the setting of economic growth rate, and the optimal resource consumption and minimum environmental cost.

3.1.3 Input-Output Model

Input-output model is a quantitative economic method to study the relationship between various sectors or products of the national economy (Leontief 1986). By compiling input-output tables and using linear algebraic tools to establish mathematical models, the model reveals the internal relations among various sectors of the national economy and reproduction links, and then carries out economic analysis, prediction and budget planning. According to different units of measurement, the model can be divided into value type and physical type.

The input-output model expresses the direct balance relationship between production and distribution in various sectors of the national economy in the input-output table in mathematical form. It can also simulate the calculation with computer software to reveal the internal relationship between the various sectors of the national economy and the various links of social reproduction. The data come from the production of the economic and social system. It is widely used in economic structure analysis, economic planning and economic development forecast, etc.

3.1.4 Research Progress

Based on the theory of material balance, value balance and energy balance, a dynamic optimal comprehensive evaluation model is established by using input-output table. Many achievements have been made in the studies on theory and application of comprehensive evaluation of economy, society and environment.

1. **Theoretical research**

The theoretical research of this evaluation model began in the 1960s, and extended to the field of natural resources and environment research through input-output model. Most scholars mainly studied how resources and pollutants affect one industry from another (Ayre 1969; Leontief 1977; Lee 1982; Forsund 1985; Perrings 1987).

In many complex economic models, physical units are used for the input of natural resources and environmental materials, while monetary units are used for economic development variables (Hannon 2001). These composite models mainly use the data information in the input-output table to predict the changes of resources input, pollutants production and economic activities in various hypothetical conditions and scenarios. It is believed that in the construction of the model, the input-output table can provide a national numerical range index and predict nutrient residues and other economic and human activities in different economic growth scenarios (Turner 2000).

Usual comprehensive evaluation models include how environmental policies affect macroeconomic and social development and the level of economic output (Moffatt and Hanley 2001). The impact of environmental policy on economy is how to distribute among different sectors of economic development and how to improve policies. The comprehensive evaluation of environmental policy based on input-output model must fully consider many factors, such as the impact of environmental policy on total output, changes in industrial structure and long-term economic growth rate, and so on.

2. **Applied research**

At the end of the twentieth century, the study of evaluation models and the introduction of theories and ideas turned to how to evaluate the economy, society and environment effectively. The dynamic optimal comprehensive evaluation model based on the "three balances" theory and input-output model is a popular tool for environmental policy evaluation, which has been applied in many countries. Some studies focus on the evaluation of GHG emission intensity policies of different countries, such as in Australia, Portugal, Korea, Greece, China and Japan (Lenzen 1998; Oliveira 2004; Chung 2009; Hristu 2010), and others focus on the evaluation of water environment improvement policies. Based on the "Three Balances" theory, some scholars have established a comprehensive evaluation model for dynamic optimization of Japan's water environment improvement policies. The model has been dynamically simulated by using computer simulation technology. An empirical study has been carried out on the basis of the second largest lake (Kasumigaura Lake) basin in Japan (Higano and Sawada 1997; Higanao and Yoneta 1999; Hirose and Higano 2000; Mizunoya 2007). Chinese scholars have constructed a dynamic model suitable for China's economic, social and environmental characteristics, which is applied to the evaluation of water environment improvement schemes (Zhang 2013; Yan 2014, 2015a, b; Xiang 2014, 2015; Xu 2015; Yang 2015a, b) and the evaluation of new energy utilization and carbon emission reduction schemes (Xu 2013; Song 2015a, b, c; Zhou 2016).

Although the dynamic optimization comprehensive evaluation model which based on the "three balances" theory and input-output model is mature in theory and widely applied in practice, there are still some problems. In studying the relationship between energy consumption and economic and social development, scholars usually neglect the changes of atmospheric environment caused by energy consumption. This book will establish a dynamic optimization comprehensive evaluation model to comprehensively evaluate the impact of haze control policies on the economic, social, atmospheric environment, pollutant emissions and energy consumption of Hebei province. On the basis of previous research results, the pollutant emission model and energy balance model are improved, and the impact of haze control policy on economy, society and environment in Hebei province is comprehensively analyzed. Policy proposals are provided for the sustainable development of economy, society and environment in Hebei province.

3.2 Conceptual Model and Basic Settings

3.2.1 Conceptual Model

The dynamic optimization comprehensive evaluation model in this book includes a target function (Maximize GRP) and three sub-models: the economic and social model of pollutant emissions model and the energy balance model (see Fig. 3.3).

Pollutant emission model describes the relationship between pollutants produced by economic and social activities. SO_2, NOx, PM_{10}, CO, VOCs, $PM_{2.5}$ and GHGes (GHG) are selected as indicators of the model. Economic model describes the relationship between capital input and social output. The Energy balance model uses standard coal as the measurement index, and the unit is ton of standard coal (TCE). It describes the balance between energy supply and demand in economic and social activities. The simulation period is 19 years from 2012 to 2030.

In this dynamic optimization comprehensive evaluation model, the three subsystems are interdependent and mutually restrictive. Economic and social activities need to consume energy while discharging pollutants into the environment. Government subsidies are introduced to improve the environment in order to achieve the goal of

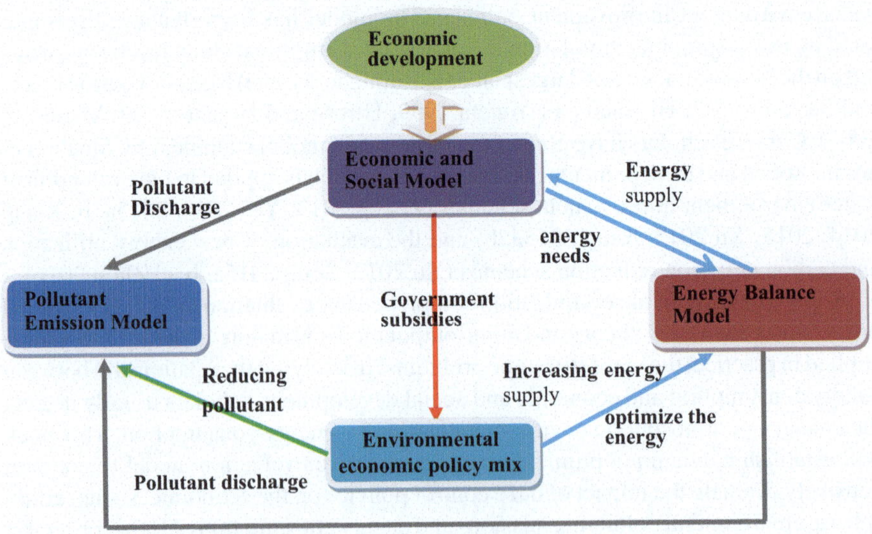

Fig. 3.3 Conceptual framework

improving the environment. Pollution environmental policy combination not only consumes energy, but also produces energy, thus affecting the energy balance.

In a word, these three sub-models and objective functions are a comprehensive system. This book models the integrated system and compiles it into a computer language for simulation experiments.

3.2.2 *Classification of Pollutants*

The main pollutant $PM_{2.5}$ of haze can be divided into primary source and secondary source. Primary source refers to the direct discharge of particulate matter into the atmosphere by pollution sources. There are both natural and human reasons in the primary source. The human reasons include industrial dust, motor vehicle exhaust particulate matter, road dust, building dust and so on. Secondary source refers to the conversion of gaseous compounds emitted by pollution sources into particulate matter through atmospheric chemical reaction process, including SO_2, NOx, NH_3, VOCs, etc. Compared with primary emission, secondary particulate matter usually has smaller particle size and a large proportion. In order to improve the environment, the methods of controlling the total amount of pollutants are used. The main components of haze and GHG are used to select pollutants. Tables 3.1 and 3.2 are the main components of haze and GHG.

Table 3.1 Classification of pollutants

Serial number	Pollutants
1	SO_2
2	NO_x
3	TPM_{10}
4	CO
5	VOCs
6	$PM_{2.5}$
7	GHG

Table 3.2 Major components of GHG

Serial number	GHG	Greenhouse effect index
1	CO_2	1
2	CH_4	25
3	N_2O	298

Table 3.3 Classification of industrial sector

Serial number	Industrial sectors
1	Agriculture
2	Construction material industry
3	Chemical industry
4	Metallurgy
5	Power
6	Other industries
7	Retail and wholesale
8	Traffic and warehousing
9	Other services

3.2.3 Preparation of Input-Output Tables

In order to better analyze the economic and environmental changes in Hebei province, this book uses the input-output table of Hebei province in 2012, and divides the input-output table into nine industrial sectors, based on the merger of industries with similar pollutant emission coefficient and energy consumption coefficient. The merged industrial sectors are shown in Table 3.3. The pollutant emission coefficient and energy consumption coefficient of various industrial sectors are calculated based on the Statistical Yearbook of Hebei province in 2012, the Environmental Situation Bulletin of Hebei province in 2012, the Water Resources Bulletin of Hebei province in 2012, the Determination of Vehicle Emission Factor of Different Emission Standards in China, the Establishment of Pollution Source Inventory of Five Provinces and Cities in North China, and the Research on countermeasures of air pollution control in Beijing. The input-output coefficient and the added value rate of each industry are shown in Appendix 1 and 2.

3.2.4 Environmental Policies and Technological Choices

In this book, environmental policy is written into the model as an exogenous variable, and the effects of different policy combinations are evaluated. The policies simulated in the model include soil and water conservation policy, industrial restructuring policy, subsidized development and utilization of clean energy policy, motor vehicle restriction policy, subsidized new energy vehicle policy and subsidized use of $PM_{2.5}$ to deal with technology and politics.

1. **Soil and water conservation policy**

Since different types of land have different emission levels of pollutants, and woodlands have almost no emissions. Therefore, the improvement of the environment can be achieved by introducing soil and water conservation policies. The policy of soil

Table 3.4 Classification of land use types

Serial number	Land use types
1	Agricultural land
2	Forestry land
3	Urban construction land
4	Other land Use

and water conservation in the model is to transform other types of land into forest land with strong soil and water conservation capacity by subsidizing the total area of land in Hebei province, so as to reduce pollutant emissions, improve the environment and reduce haze, and achieve sustainable and coordinated development of environment and economy. Table 3.4 shows the classification of land use types in Hebei province.

2. **Industrial structure adjustment policy**

Under the premise of maximizing economic development, the policy of industrial restructuring in the model uses Harrod-Doma model to describe the capital-output relationship after industrial subsidies. Through government subsidies, the capital stock of high-polluting and energy-consuming industries can be reduced, so as to reduce the production capacity of such enterprises, reduce environmental pollution and achieve the goal of harmonious development of environment and economy.

3. **Subsidizing the development and utilization of new energy policies**

The policy of subsidizing the development and utilization of clean energy is that the government uses wind and solar power generation technology through financial subsidies, increases the use of wind and solar energy to replace traditional energy, optimizes the energy utilization structure, and achieves the purpose of haze reduction. The technical parameters of model wind and solar energy are shown in Table 3.5.

4. **Vehicle restriction policy**

Vehicle restriction policy is to reduce the number of vehicles on road by restricting the traffic of all kinds of vehicles, so as to effectively reduce vehicle exhaust emissions

Table 3.5 Technical parameters of wind and solar power generation

Technology	Type	Investment cost (10,000 RMB Yuan)	Generation cost (RMB/kwh)	Operation and maintenance cost (10,000 RMB Yuan/ton standard coal)	Expected grid price (RMB/kWh)
Solar energy	5 KW	4.5	2.0	0.00895	1.00
	20 MW	19,000	1.8	0.00813	1.00
Wind energy	49.5 MW	42,500	0.5	0.00707	0.57
	100 MW	90,000	0.5	0.00659	0.77–0.98

and improve air quality. At present, the restriction policy of motor vehicle in Hebei province is limited to two tail numbers and single or double numbers. According to the data published by Hebei Environmental Protection Department and the actual survey, the number of restricted days and the rate of restricted traffic (the ratio of restricted vehicles to all vehicles) are calculated. In the model, car uses are limited to 80 days per year, and the rate of restricted traffic is 80%.

5. **Subsidy policy for new energy vehicles**

The policy of subsidizing new energy automobiles is to achieve the purpose of alleviating air pollution and reducing haze emission by increasing the use of new energy automobiles through government subsidies for zero-emission new energy automobiles and reducing the annual growth of traditional automobiles on the premise of meeting the market demand for automobiles.

6. **Subsidies of technical against $PM_{2.5}$ policies**

The policy of subsidizing the $PM_{2.5}$ treatment technology is to encourage enterprises to use advanced $PM_{2.5}$ treatment technology to reduce pollutant emissions through financial subsidies. The $PM_{2.5}$ treatment technology policies introduced in the model include bag dust removal technology (bag dust removal method), combined bag dust removal technology (electrostatic dust removal method), wet electrostatic dust removal technology (wet dust condensation method) in coal-fired power plants and biofilm dust suppression technology (bio-nano dust suppression method) which are supplemented by the government to increase the use of dust removal technology, end-treatment of pollution. Table 3.6 gives details of the $PM_{2.5}$ treatment technologies.

Table 3.6 Details of $PM_{2.5}$ treatment technologies

$PM_{2.5}$ treatment technology	Cost of investment (10,000 RMB Yuan)	Operating expenses (10,000 RMB Yuan)	Annual dust removal capacity (tons)	Dust removal rate (%)
Cloth bag dust removal	2200	120	1890	99.9
Unit electric bag composite dust removal	2750	400	700	99.1
Wet electrostatic precipitation in coal-fired power plants	2500	312	1052.7	90
Biological nano film for dust suppression	2000	60	1800	99

3.3 Modeling

The comprehensive evaluation model for dynamic optimization of fog and haze control policies in Hebei province consists of an objective function and three sub-models. The objective function is to maximize the GRP. The three sub-models are economic and social model, pollutant emission model and energy balance model. The simulation period is 19 years from 2012 to 2030. The variables in the model are divided into endogenous variables (en) and exogenous variables (ex).

3.3.1 *Objective Function*

The objective function of the model is to maximize the GRP. According to the utility maximization theory, the objective function makes the economic and social system and the market reach equilibrium state. The objective function is expressed by the following equation.

$$\text{Max} \sum_{t} \frac{1}{(1+\rho)^{(t-1)}} GRP(t) \ (\text{t} = 1, 2 \ldots, 11) \tag{3.1}$$

$$GRP(t) = \sum_{m} v^{m} \cdot X^{m}(t) \ (m = 1, 2 \ldots, 9) \tag{3.2}$$

The variables in Eqs. (3.1) and (3.2) are described as follows.

t	In the simulation period, the values are 1–19;
m	Industrial sector, with a value of 1–9;
ρ	Social depreciation rate, 0.05 (ex);
$GRP(t)$	GRP of Hebei province at the t-th period (en);
v^{m}	The value-added rate of industrial sector m (ex);
$X^{\mathrm{m}}(t)$	The output of the industrial sector m in Hebei province in the t-th period (en).

3.3.2 *Pollutant Emission Model*

The structure of pollutant emission model in Hebei province is shown in Fig. 3.4. Pollutant emissions include three sources: point source pollution, non-point source pollution and mobile source pollution.

Point source pollution is air, water, heat, noise or light pollution caused by identifiable single source of pollution. The discharge of point source pollutants includes pollution from residents' living and consumption, industrial production activities and service activities. Pollution from industrial production activities includes pollution

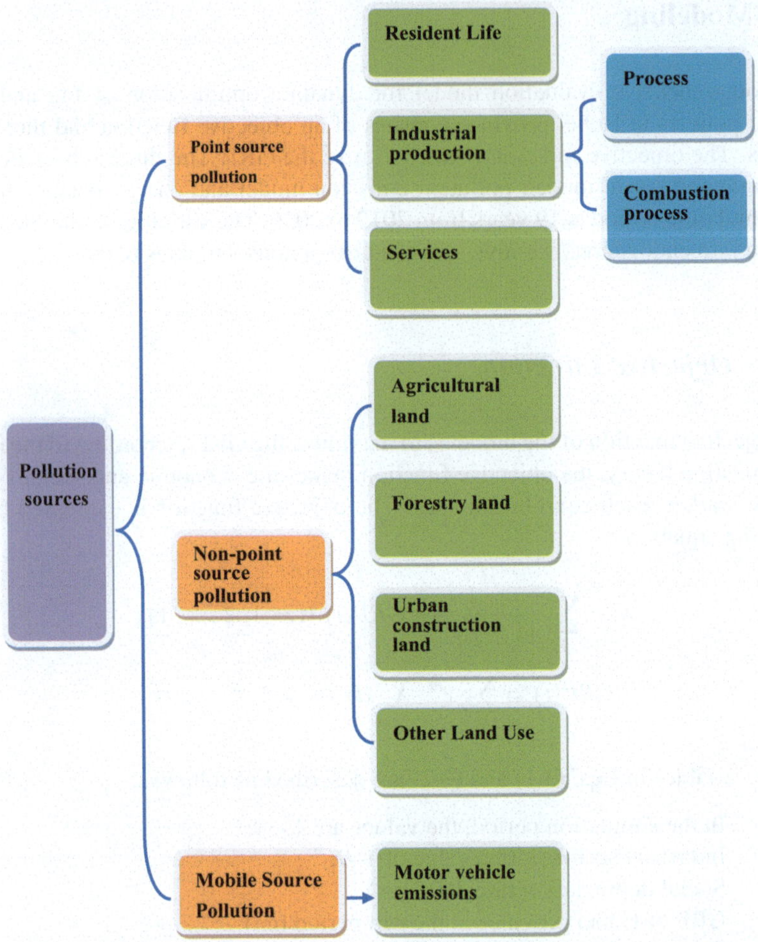

Fig. 3.4 Pollutant discharge model

from industrial combustion processes and technological processes. The introduced haze treatment technology is applied to point source pollution.

Diffused Pollution (DP), also known as Non-point Source Pollution (NPS), refers to the discharge of dissolved and solid pollutants from non-specific sites. Non-point source pollution mainly refers to the pollutants discharged from various types of land.

Mobile source pollution is a spatial pollution source whose position changes with time and is from vehicle exhaust. It mainly refers to vehicles that emit air pollutants, such as carbon oxides, nitrogen oxides, sulfur oxides, hydrocarbons, lead compounds and black smoke, including automobiles, aircraft, ships, locomotives, etc. vehicle exhaust pollution.

1. **Total Emission of pollutants in Hebei province**

The total amount of pollutant discharge in Hebei province is the sum of the total pollutant load of each pollution source. It's expressed by the following equation:

$$TQ^p(t) = NQ^p(t) + PSP^p(t) + MSP^p(t) \tag{3.3}$$

($p = 1$: SO_2; $p = 2$: NO_x; $p = 3$: TPM_{10}; $p = 4$: CO; $p = 5$: VOCs; $p = 6$: $PM_{2.5}$; $p = 7$: GHG)

In Eq. (3.3), the variables are described as follows.

- $TQ^p(t)$ Total amount of pollutant p discharged from Hebei province in the t-th period (en);
- $NQ^p(t)$ Emission of pollutant p from non-point source pollution in the t-th period (en);
- $PSP^p(t)$ Emission of pollutant p from point source pollution in the t-th period (en);
- $MSP^p(t)$ Emission of pollutant P from mobile source pollution in the t-th period (en).

2. **Emissions of non-point source pollutants**

The pollutants discharged from non-point source pollution are determined by land use types and their pollutant emission coefficients. The pollutant p emission coefficients of land use type G are calculated based on the data from the pollution source emission inventory of Chengde City in 2013 and the land area of Chengde City, as shown in Appendix 3.

The equation of non-point source pollution is as follows:

$$NQ^p(t) = \sum_g EL_g^p \cdot L^g(t) \tag{3.4}$$

In Eq. (3.4), the variables are described as follows.

- $NQ^p(t)$ Emission of pollutant p from non-point source pollution in the t-th period (en);
- EL_g^p Emission coefficients of pollutant p in land use type g (ex);
- $L^g(t)$ The area of land use type g (en).

3. **Emission of point source pollutants**

The emission of point source pollutants is the pollution caused by household consumption, industrial production activities and service industry activities minus the amount of $PM_{2.5}$ treated by introducing $PM_{2.5}$ treatment technologies, and the annual $PM_{2.5}$ treatment volume is set to be no more than 80% of the current year's point source. The data and coefficient of dust removal technology are calculated through the data in the compilation of advanced technologies for air pollution prevention and control, as shown in Appendix 4. The pollution from industrial production activities includes the pollution from industrial combustion process and technological process. The pollutant emission from industrial production depends on the pollutant emission coefficient of output value and unit output value, see Appendix 5–7. The emission of pollutants from residents' living activities is determined by the final consumption and the emission coefficient per unit of output value of pollutants P, as shown in Appendix 8 and 9, and transportation industry is not included in service industry and residents' living consumption. The emission coefficient of pollutant p of industry m is calculated according to the data in "Establishment and Uncertainty Analysis of Pollution Source Inventory in Five Provinces and Cities of North China" and "Hebei Economic Yearbook"; The emission coefficient of pollutant p unit output value is calculated based on the data of Hebei Statistical Bureau; The growth rate of permanent population in Hebei province is the root. The population of Hebei province in recent years was calculated according to Hebei Economic Yearbook. The equation is described as follows:

$$PSP^p(t) = UIQ^p(t) + HQ^p(t) + FWQ^p(t) - \sum_{o} RE^o(t) \tag{3.5}$$

$$UIQ^p(t) = \sum_{m} X^m(t) \cdot EUI_m^p + \sum_{m} X^m \cdot PUI_m^p (m = 2, 3, 4, 5, 6) \tag{3.6}$$

$$HQ^p(t) = C^m(t) \cdot EH^p (m \neq 8) \tag{3.7}$$

$$FWQ^p(t) = X^m(t) \cdot FH^p (m = 7, 9) \tag{3.8}$$

$$C^m(t+1) = \omega^m \cdot Z(t+1) \tag{3.9}$$

$$Z(t+1) = Z(t) \cdot (1 + \mu) \tag{3.10}$$

The variables in Eqs. (3.5)–(3.10), are described as follows.

$UIQ^p(t)$	Emission of pollutant p produced by industrial production in the t-th period(en);
$HQ^p(t)$	Emission of pollutant p from residents' lives in the t-th period (en);
$FWQ^p(t)$	Emission of pollutant p from service industry activities in the t-th period (en);
$RE^o(t)$	$PM_{2.5}$ removal of dust removal technology o in the t-th period (en);
$C^m(t)$	Final consumption of industry m in the t-th period (en);
ω^m	Per capita consumption coefficient of industry m (ex);
$Z(t)$	Population of Hebei province in the t-th period (en);
EUI_m^p	Emission coefficients of pollutant p in combustion process industry m (ex);
PUI_m^p	Emission coefficient of pollutant p in process industry m (ex);
EH^p	Discharge coefficient of final consumption pollutant p per unit output value (ex);

FH^p	Discharge coefficient of pollutant p per unit output value in service industry (ex);
μ	The growth rate of the permanent population in Hebei province (ex).

4. **Emissions of pollutants from mobile sources**

Pollutants from mobile sources include emissions of cars, agricultural vehicles, motorcycles and new energy vehicles. This book assumes that cars and motorcycles use gasoline, new energy vehicles use electricity, agricultural vehicles use diesel, and motor vehicles shall adopt State iii standards. The emission coefficient of pollutant P of motor vehicle type C is based on the data of "Determination of Vehicle Emission Factor of Different Emission Standards in China", see Appendix 10; the equation is described as follows (C = 1: car; C = 2: agricultural vehicles; C = 3: motorcycles; C = 4: new energy sources vehicles):

$$MSP^p(t) = TMV^c(t) \cdot \zeta_c^p \tag{3.11}$$

In Eq. 3.11,

$MSP^p(t)$	Emission of pollutant p from mobile source pollution in the t period (en);
ζ_c^p	Emission coefficient of pollutant p of motor vehicle type c (ex);
$TMV^c(t)$	Number of types of motor vehicles in the t period (en).

3.3.3 Energy Balance Model

The energy model describes the relationship between energy supply and demand. The conceptual model of energy balance is shown in Fig. 3.5. Energy demand is determined by industrial production and final consumption. Energy supply comes

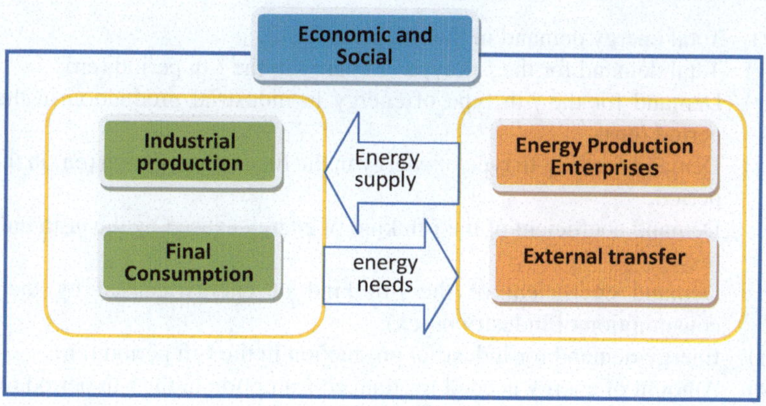

Fig. 3.5 Energy balance model

from energy production enterprises and external transfer. In order to achieve economic and social development, the total energy supply in the model is supposed to greater than or equal to the total energy demand.

1. **Total energy demand**

Energy demand consists of two parts: industrial production and final consumption. The consumption coefficient per capita, the demand coefficient of the *j*-th type of energy caused by industry and final consumption are calculated based on the data of Hebei Statistical Yearbook and the National Statistical Bureau, see Appendix 11 and 12. The total energy demand can be adjusted and controlled by means of adjusting industrial structure and motor vehicle restriction policy. The total energy demand is divided into four categories: coal, oil, natural gas and primary power (new energy). The total energy demand can be expressed by the following equation:

$$TED(t) = \sum_{j} TED_j(t)(j = 1, 2, 3, 4) \tag{3.12}$$

$$TED_j(t) = PED_j(t) + FED_j(t) \tag{3.13}$$

$$PED_j(t) = \sum_{m} X^m(t) \cdot ED_j^m \tag{3.14}$$

$$FED_j(t) = \sum_{m} C^m(t) \cdot FEC_j^m \tag{3.15}$$

$$PED(t) = \sum_{j} PED_j(t) \tag{3.16}$$

$$FED(t) = \sum_{j} FED_j(t) \tag{3.17}$$

The variables in Eqs. (3.12)–(3.17) are described as follows.

$TED(t)$	Total energy demand in the t-th period (en);
$TED_j(t)$	Total demand for the j-th type of energy in the t-th period (en);
$PED_j(t)$	Demand for the j-th type of energy in industrial production in the t-th period (en);
$FED_j(t)$	Demand for the j-th type of energy in the final consumption (en) in the t-th period;
ED_j^m	Demand coefficient of the j-th kind of energy caused by the m-th industry (ex);
FEC_j^m	Demand coefficient of the j-th kind of energy caused by the final consumption of industry m (ex);
$PED(t)$	Energy demand for industrial production in the t-th period (en);
$FED(t)$	Amount of energy needed for final consumption in the t-th period (en);
$X^m(t)$	Output of the industrial sector m in Hebei province in the t-th period (en);
$C^m(t)$	Final consumption of industry m in the t-th period (en).

2. **Total energy supply**

In the model, the total energy supply of Hebei province includes coal, oil natural gas and primary power. It is assumed that the increase of primary energy supply is the increase of wind and solar power generation. The annual growth rate of various energy sources is calculated according to the energy planning of Hebei Statistical Bureau and Hebei province. The equation is described as follows:

$$TES(t) = \sum_j TES_j(t) \tag{3.18}$$

$$TES_j(t) = W^q(t) + TE^{\mathrm{d}}(t)(j = 4) \tag{3.19}$$

$$TES_j(t+1) = TES_j(t) \cdot \partial_j \tag{3.20}$$

The variables in Eqs. (3.18)–(3.20) are described as follows.

q	Wind power generation technology type, value 1, 2 (ex);
d	Solar power generation technology type, value 3, 4 (ex);
$TES(t)$	The total amount of energy supply in the t period (en);
$TES_j(t)$	The j-th kind of energy supply in the t period (en);
∂_j	Annual growth rate of energy supply of type j (ex);
$W^q(t)$	The total amount of wind power generation technology q in the t period (en);
$TE^{\mathrm{d}}(t)$	The total amount of solar power generation technology d in the t period (en).

3. **Energy balance**

In order to ensure the normal operation of economic and social activities, energy supply is set to be equal to or greater than energy demand. It is described by the following equation:

$$TED_j(t) \leq TES_j(t) \tag{3.21}$$

3.3.4 Economic and Social Model

In order to realize the sustainable development of environment, economy and society in Hebei province, the model simulates six environmental improvement policies, namely, soil and water conservation policy, industrial structure adjustment policy, motor vehicle restriction policy, subsidizing new energy vehicle policy, introducing $PM_{2.5}$ treatment technology policy and subsidizing the development and utilization of clean energy policy. In this section, the environmental improvement policy of Hebei province and its impact on industrial production and market balance are modeled. After the simulation experiments, the environmental and economic impact of haze control policy in different scenarios is analyzed and compared.

1. **Soil and water conservation policy**

Soil and water conservation policies are used to improve land use, increase forestry area, enhance soil and water conservation capacity and reduce non-point source pollution. Through financial subsidies, other land use types are transformed into woodlands to improve the quality of water and environment, and reduce haze. The equation is described as follows:

$$\bar{L}(t) = \sum_{g} L^{g}(t) \tag{3.22}$$

In the equation, the variables are described as follows.

$\bar{L}(t)$ Total land area of Hebei province in the t-th period (en);
$L^{g}(t)$ The area of land type g in Hebei province in the t-th period (en);

The forest land area changes with the government subsidy, and the financial subsidy coefficient is the average value calculated according to the data in the New Standard of Land Expropriation Compensation in Hebei province 2016. The equation is described as follows:

$$L^{2}(t+1) = L^{2}(t) + \Delta L^{2}(t)(g = 2) \tag{3.23}$$

$$\Delta L^{2}(t) = L^{42}(t) \tag{3.24}$$

$$L^{42}(t) \geq \lambda^{4} \cdot TS(t) \tag{3.25}$$

The variables in Eqs. (3.22)–(3.25) are described as follows.

$L^{2}(t)$ Forest area in the t-th period (en);
$\Delta L^{2}(t)$ The increased forest area in the t-th period (en);
$L^{42}(t)$ The land area of forest (g = 2) was changed from other land use types (g = 4) in the t-th period (en);
λ^{4} The reciprocal coefficient of land financial subsidy (ex);
$TS(t)$ Financial amount of afforestation subsidized by Hebei Provincial Government in the t-th period (en).

2. **Industrial structure adjustment policy**

Through the adjustment of the industrial structure to achieve the goal of reducing the emission of pollutants. Harrod-Domar model is used to describe the capital-output relationship after industrial subsidies. The capital input-output ratio of industry M is calculated according to the input-output table of Hebei province in 2012, see

Appendix 13 and 14. The capital depreciation rate of industry m is uniformly 0.1. The equation is as follows:

$$X^m(t) \le \alpha^m\{k^m(t) - S^m(t)\}\ (m = 1, 2, \ldots, 15) \tag{3.26}$$

$$k^m(t+1) = k^m(t) + I^m(t+1) - f^m \cdot k^m(t) \tag{3.27}$$

Variables in Eqs. (3.26) and (3.27) are described as follows:

$X^m(t)$	Output of the industrial sector m in Hebei province in the t-th period (en);
$k^m(t)$	The actual capital of industry m in the t-th period (en);
$S^m(t)$	The government's financial subsidy to industry m in the t-th period (en);
α^m	Capital input-output ratio of industry m is reciprocal (ex);
$I^m(t+1)$	Investment in industrial sector m in the $(t+1)$ t-th period (en);
f^m	Capital depreciation rate of industry m (ex)

3. **Motor vehicle restriction policy**

By introducing the implementation of the motor vehicle restriction policy, the frequency of vehicle use could be declined and the goal of reducing the emission of pollutants could be achieved. The motor vehicle motor vehicle restriction policy and the limit rate of the use of cars are collected based on Hebei provincial environmental protection department and actual survey data. The equation is described as follows:

$$DMSP^p(t) = \sum_c \left[(1-\mu) \cdot TMV^c(t) \cdot \zeta_c^p\right](c = 1) \tag{3.28}$$

The variables in Eq. (3.28) are described as follows:

$DMSP^p(t)$	The emission of pollutant p reduced by the implementation of the motor vehicle restriction policy in the t-th period (en);
μ	Car restriction rate (ex);
ζ_c^p	The emission coefficient of pollutant p of motor vehicle type C (ex);
$TMV^c(t)$	The number of types of motor vehicles in the t-th period (en).

4. **Subsidies for new energy vehicles**

The policy of subsidizing new energy vehicles is realized by increasing the use of new energy vehicles. The subsidy rate of new energy vehicles is the average value calculated according to the national regulations; It is assumed in this research that the added new energy vehicles are cars. The equation is described as follows:

$$TMV^8(t+1) = TMV^8(t) + DTMV^8(t) \tag{3.29}$$

$$NS(t) = \chi \cdot DTMV^8(t) \tag{3.30}$$

Variables follows in Eqs. (3.29) and (3.30) are described as:

$DTMV^8(t)$ The annual increase of new energy vehicles in the t-th period (en);
$NS(t)$ The government subsidies for new energy automobiles in the t-th period (en);
χ Subsidies for new energy vehicles (ex).

5. **Policy of subsidizing the development and utilization of clean energy**

The policy of subsidizing the development and utilization of clean energy is realized by subsidizing the development and utilization of wind and solar energy technologies. Detailed technical parameters of wind and solar energy are listed in Appendix 15. The equation is described as follows:

$$W^{q}(t+1) = W^{q}(t) + \Delta W^{q}(t+1) \tag{3.31}$$

$$\Delta W^{q}(t+1) = WI^{q}(t) \cdot \theta^{q} (\mathrm{q} = 1, 2) \tag{3.32}$$

$$TE^{d}(t+1) = TE^{d}(t) + \Delta TE^{d}(t+1) \tag{3.33}$$

$$\Delta TE^{d}(t+1) = EI^{d}(t) \cdot \varpi^{d} (p = 1, 2) \tag{3.34}$$

$$WS^{q}(t) = WI^{q}(t) + WM^{q}(t) \tag{3.35}$$

$$WM^{q}(t) = W^{q}(t) \cdot \phi^{q} \tag{3.36}$$

$$ES^{d}(t) = EI^{d}(t) + EM^{d}(t) \tag{3.37}$$

$$EM^{d}(\mathrm{t}) = TE^{d}(t) \cdot \phi^{d} \tag{3.38}$$

The variables in Eqs. (3.31)–(3.38) are described as follows.

$WI^{q}(t)$ The investment amount of new wind power generation technology q in the t-th period (en);
θ^{q} Investment conversion coefficient of wind power generation technology q (ex);
$\Delta TE^{d}(t+1)$ Increased electricity from solar power technology d in the (t + 1)-th period (en);
$EI^{d}(t)$ The investment in solar power generation technology d in the t-th period (en);
ϖ^{d} Investment conversion factor of solar power technology d (ex);
$WM^{q}(t)$ The management cost of operation and maintenance of wind power generation technology q in the t-th period (en);
$EM^{d}(t)$ The management cost of operation and maintenance of solar power generation technology d in the t-th period (en);

ϕ^q	Management Cost Conversion Coefficient of wind power technology q (ex);
ϕ^{d}	The management cost conversion coefficient of solar power technology d (ex);
$WS^{\mathrm{q}}(t)$	Government subsidy for the advanced wind power technology q in the t-th period (en);
$ES^{\mathrm{d}}(t)$	Government subsidy for the advanced solar power technology d in the t-th period (en).

6. **$PM_{2.5}$ treatment technology policy**

$PM_{2.5}$ treatment technology policy is achieved by subsidizing the introduction of bag dust removal technology (bag dust removal method), unit electric bag composite dust removal technology (electrostatic dust removal method), coal-fired power plant wet electrostatic dust removal technology (wet dust condensation method) and bio-nano dust suppression technology (bio-nano dust suppression of Bermeticam). It is set in this book that the annual $PM_{2.5}$ treatment volume should not exceed 80% of the current year's point source. The data and coefficients of the dust removal technology are obtained from the data in the Compilation of Advanced Technologies for the Prevention and Control of Atmospheric Pollution. The equation is as follows:

$$RE^o(t+1) = RE^o(t) + \Delta RE^o(t+1) \tag{3.39}$$

$$RE^o(t) \leq 0.8 \cdot PSP^p(t)(p = 6) \tag{3.40}$$

$$\Delta RE^o(t+1) = RI^o(t) \cdot RB^o(k = 1, 2, 3, 4) \tag{3.41}$$

$$RM^o(t) = RE^o(t) \cdot \sigma^o \tag{3.42}$$

$$RS^o(t) = RI^o(t) + RM^o(t) \tag{3.43}$$

The variables in Eqs. (3.39)–(3.43) are described as follows.

$\Delta RE^o(t+1)$	Reduced $PM_{2.5}$ emissions from dust removal technology o in the (t + 1)-th period (en);
$RI^o(t)$	The investment of advanced dust removal technology o in the t-th period (en);
RB^o	Investment conversion coefficient of dedusting technology o (ex);
$RM^o(t)$	The management cost of operation and maintenance of dust removal technology o in the t-th period (en);
σ^o	The management cost conversion coefficient of dedusting technology o (ex);
$RS^o(t)$	Government subsidy for dust removal technology o in the t-th period (en).

7. **Government subsidies**

Government subsidy for air environment improvement policy includes government subsidy for afforestation, industrial restructuring, motor vehicle restriction, new energy vehicles and management expenses for early investment and later operation and maintenance of new energy. The financial budget allocated to each policy should not exceed the government's total budget for the environmental improvement policies of the year. The equation is as follows:

$$GS(t) \geq TS(t) + \sum_{m} S^{m}(t) + NS(t) + \sum_{q} WS^{q}(t) + \sum_{d} SES^{\mathrm{d}}(t) + \sum_{o} RS^{o}(t) \tag{3.44}$$

The variables in Eq. (3.44) are described as follows.

$GS(t)$ The total financial subsidy for the air environment improvement policies of Hebei province in the t-th period (en).

8. **Market balance of industrial production**

Input-output model is the basis of research. According to market equilibrium, the total output of each industry ought to be greater than or equal to the sum of final demand and intermediate input. The input-output coefficient of the n-th industry is calculated from the input-output table of Hebei province in 2012. The equation of dynamic equilibrium relationship of industrial development is described as follows:

$$X^{m}(t) \geq Q^{m} \cdot X^{m}(t) + C^{m}(t) + i^{m}(t) + e(t) \tag{3.45}$$

$$X(t) = \sum_{m} x^{m}(t) \tag{3.46}$$

Variables in Eqs. (3.45) and (3.46) are described as follows:

Q^{m} Input-output coefficient of the m-th industry (ex);
$C^{m}(t)$ The final consumption of industry m in the t-th period (en).
$i^{m}(t)$ Total investment in the m-th industry in the t-th period (en);
$e(t)$ Total net exports in the t-th period (en).

3.4 Summary

This chapter explains the dynamic optimization comprehensive evaluation model of environmental and economic impact of haze control policy in Hebei province thoroughly. Based on the "three balances" theory, the model uses input-output table and multi-objective linear optimization method to analyze the impact of comprehensive environmental improvement policy on economy, society and environment.

Chapter 4
Scenario Analysis of Comprehensive Impact of Haze Control Policy on Economy, Society and Environment in Hebei Province

The simulation results are analyzed in this section. Firstly, the scenario design is introduced. Secondly, by comparing the simulation results under different scenarios, the optimal scenarios are found out by using environmental indicators, economic indicators, energy indicators, etc. Finally, the environmental and economic impact of haze control in Hebei province under the optimal scenario is analyzed.

4.1 Scenario Setting

This chapter sets up six scenarios to analyze the impact of the implementation of haze control policy on the environment and economy in Hebei province, and to find out the best scenario that is in line with the sustainable development of the economy, society and environment, so as to provide scientific policy recommendations to the Hebei Provincial Government.

According to the emission reduction rate of pollutants in the 12th Five-Year Plan of Environmental Protection and the 13th Five-Year Plan of Environmental Protection, the limitation set in this book is that $PM_{2.5}$ emission by 2030 will be 30% lower than that in 2016, The haze control policies adopted are soil and water conservation policy and industrial structure adjustment policy, subsidies for new energy vehicles, development and utilization of clean energy, motor vehicle restriction policies and subsidies to introduce $PM_{2.5}$ treatment technology policy.

This section sets six scenarios based on policy effectiveness and technology types, as shown in Table 4.1.

The basic scenario is set to reduce $PM_{2.5}$ emission by 30% in 2030 compared with 2016, and adopt the policy of industrial restructuring with subsidies.

The improvement scenario 1 is set to reduce $PM_{2.5}$ emission by 30% in 2030 compared with 2016, and adopt subsidized industrial structure adjustment policy and subsidized soil and water conservation policy.

G. Zhang et al., *The Effect Evaluation of Haze Governance Policies in Hebei Province–Based on I-O Model*, https://doi.org/10.1007/978-981-15-3757-8_4

Table 4.1 Scenario settings

Scenario setting	$PM_{2.5}$ emission reduction rate (%)	Subsidized soil and water conservation	Subsidizing industrial structure adjustment	Subsidies for new energy vehicles	Subsidizing the development and utilization of clean energy	Vehicle restriction	Subsidized $PM_{2.5}$ treatment technology
Basic scenario	30	No	Yes	No	No	No	No
Improvement scenario 1	30	Yes	Yes	No	No	No	No
Improvement scenario 2	30	No	Yes	Yes	No	No	No
Improvement scenario 3	30	No	Yes	No	Yes	No	No
Improvement scenario 4	30	No	Yes	No	No	Yes	No
Comprehensive scenario	30	Yes	Yes	Yes	Yes	Yes	Yes

The improvement Scenario 2 is set to reduce $PM_{2.5}$ emission by 30% by 2030 compared with 2016, and adopt policies to subsidize industrial restructuring and new energy vehicles.

The improvement Scenario 3 is set to reduce $PM_{2.5}$emission by 30% by 2030 compared with 2016, and adopt policies to subsidize industrial restructuring and the development and utilization of clean energy.

Improvement scenario 4, set as the total $PM_{2.5}$ emission in 2030 will be reduced by 30% compared with 2016, adopting subsidized industrial restructuring policy and motor vehicle restriction policy.

The comprehensive scenario is set to reduce $PM_{2.5}$ emission by 30% in 2030 compared with 2016. The policies of subsidizing industrial restructuring, subsidizing soil and water conservation, subsidizing new energy vehicles, subsidizing the development and utilization of clean energy, and introducing $PM_{2.5}$ treatment technology into motor vehicle restriction and subsidies are adopted.

4.2 Optimal Scenario Selection

This section chooses the total amount, change and average growth rate of GRP in the simulation period to analyze the economic development of Hebei province in each scenario. The total amount of pollutants discharged and the intensity of pollutants discharged in the simulation period are chose to analyze the environmental improvement of Hebei province in each scenario. The total amount of energy utilization and the intensity of energy consumption in the simulation period are chose to analyze the situation of energy utilization in Hebei province in each scenario. Finally, by synthetically comparing the economic, social, energy and pollutant indicators in different scenarios, the optimal scenario which can achieve the sustainable development of economy, energy and environment of Hebei province is selected.

4.2.1 Economic Development Analysis

1. Analysis of total economic quantity in the simulated period

The GRP of Hebei province in each scenario from 2012 to 2030 is shown in Fig. 4.1. The total amount of GRP in basic scenario is 72.19 trillion RMB Yuan, while that in improved scenarios 1, 2, 3, and 4, and comprehensive scenario is 77.31 trillion RMB Yuan, 72.20 trillion RMB Yuan, 75.12 trillion RMB Yuan, 72.20 trillion RMB Yuan, and 114.21 trillion RMB Yuan, respectively.

In the basic scenario where only the industrial restructuring policy is adopted, the total GRP is the lowest under the same limit of total $PM_{2.5}$ emission.

In the improvement scenario 1 where industrial restructuring and soil and water conservation policies were adopted simultaneously. The total GRP increased by

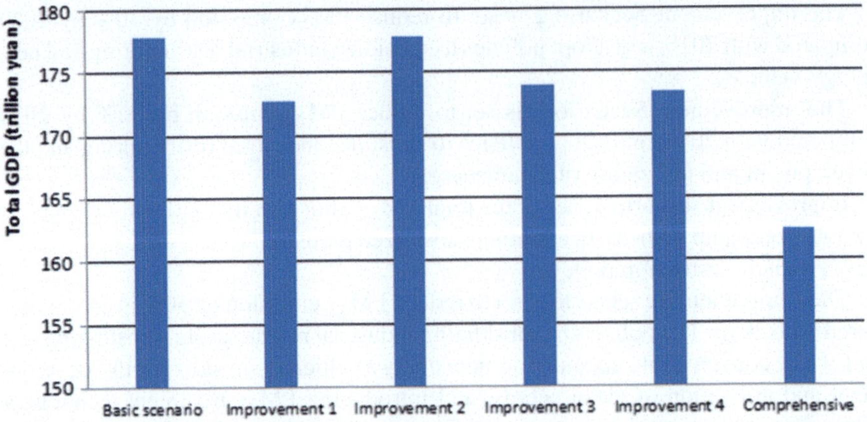

Fig. 4.1 Total GRP of Hebei province in the simulated period

5.12 trillion RMB Yuan under the same limit of total $PM_{2.5}$ emission compared with the basic scenario. It can be concluded that soil and water conservation policy is conducive to the sustainable development of environment and economy in Hebei province.

In the improvement scenario 2, two policies of industrial restructuring and subsidizing new energy vehicles are adopted simultaneously. Under the same limit of total $PM_{2.5}$ emission, the total economic output increased by 0.01 trillion RMB Yuan compared with the basic scenario. However, the total GRP of the improvement scenario 2 was lower than that of the improvement scenario 1. It indicates that although the policy of subsidizing new energy vehicles is conducive to the sustainable development of environment and economy in Hebei province, its effect is inferior to that of soil and water conservation policy.

In the improvement scenario 3, both industrial restructuring policy and subsidy policy are adopted to develop and utilize clean energy. Under the same limit of total $PM_{2.5}$ emission, the total GRP increases by 2.93 trillion RMB Yuan compared with the basic scenario, and the total economic output of the improvement scenario 3 is lower than that of the improvement scenario 1, but higher than that of the improvement scenario 2. It demonstrates that, on the premise of maintaining the sustainable development of environment and economy in Hebei province, the effect of the policy of subsidizing the development and utilization of clean energy is better than that of subsidizing new energy vehicles, but the effect is slightly worse than that of the policy of soil and water conservation.

In the improvement scenario 4, both industrial restructuring policy and motor vehicle restriction policy are adopted. Under the same limit of total $PM_{2.5}$ emission, the total GRP increases by 0.01 trillion RMB Yuan compared with the basic scenario, and the total economic output of the improvement scenario 4 is lower than that of the improvement scenario 1 and the improvement scenario 3. It suggests that the effect of motor vehicle restriction policy in realizing sustainable development of environment and economy in Hebei province is slightly worse than that of water and soil conservation and subsidized development and utilization of clean energy policy.

In the comprehensive scenario, the policies of industrial restructuring, soil and water conservation, subsidizing the development and utilization of clean energy, subsidizing new energy vehicles and subsidizing motor vehicle restriction are adopted, and $PM_{2.5}$ treatment technology are introduced. Under the same limit of total $PM_{2.5}$ emissions, the total GRP of the comprehensive scenario is the highest, which is 42.02 trillion RMB Yuan more than that in the basic scenario and other scenarios, respectively. It reveals that the comprehensive scenario of adopting six kinds of environmental policies is easier to achieve the goal of sustainable development of environmental economy, and the effect is dramatic.

From the perspective of policy effect, the policies of subsidy for soil and water conservation, subsidy for development and utilization of clean energy, subsidy for new energy vehicles, motor vehicle restriction and subsidy for introduction of $PM_{2.5}$ treatment technology have diminishing effects. Comprehensive policy can better achieve the goal of sustainable development of economy, society, energy and environment than single policy, and the effect of "source governance" policy is superior than that of "end governance" policy.

2. Analysis of GRP variation trend

The variation trend of GRP in Hebei province from 2012 to 2030 is shown in Fig. 4.2. The GRP of basic scenario, improved scenarios 1, 2, 3, and 4, and comprehensive scenario all increased first and then decreased, with the inflection point appearing around 2021. In the six scenarios, the average annual growth rates are 1.38%, 2.56%, 1.38, 1.87%, 1.38% and 5.87% respectively. The simulation results show that all kinds of environmental policies can promote economic development under the same

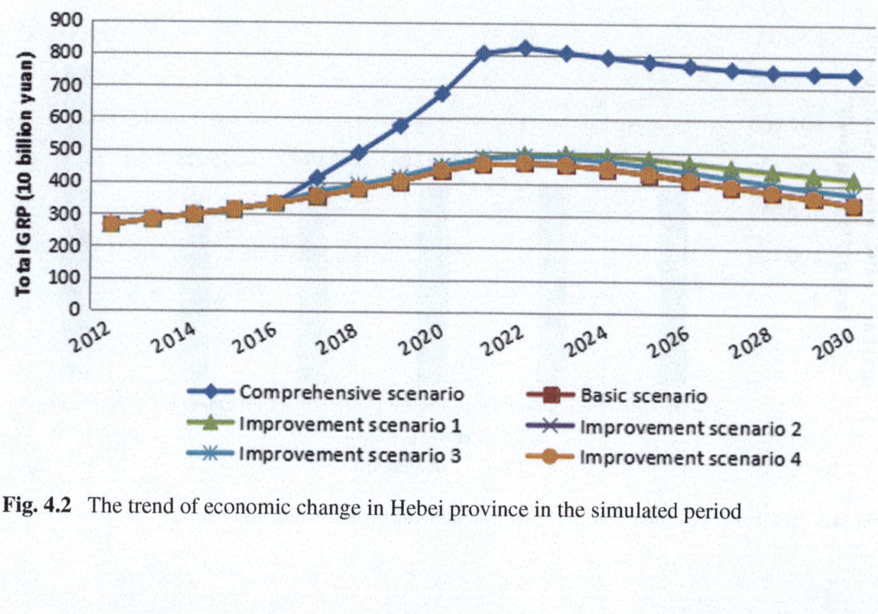

Fig. 4.2 The trend of economic change in Hebei province in the simulated period

limit of total $PM_{2.5}$ emission, but it is difficult for Hebei province to maintain high-speed economic growth under the constraint of total emission reduction. From 2017 to 2021, with the implementation of various policies and the introduction of new technologies, pollutant emissions declined rapidly and the policy effect is excellent. After 2022, the adjustment of industrial structure and energy structure has reached or approached the upper limit of control. The emission reduction of haze treatment technology has become a "bottleneck" and Hebei province has to reduce the total economic output to achieve the goal of environmental emission reduction.

4.2.2 *Energy Consumption Analysis*

1. Analysis of total energy and energy intensity

The total energy use and intensity of each scenario during the simulation period are shown in Figs. 4.3 and 4.4, respectively. The total energy utilization in basic scenario, improvement scenarios 1, 2, 3, and 4 is 4501.75 million tons standard coal at minimum, 4632.83 million tons of standard coal, 4501.88 million tons of standard coal, 4616.80 million tons of standard coal, 4501.87 million tons of standard coal, respectively. The energy utilization intensity in basic scenario, improvement scenarios 1, 2, 3, and 4 is 0.6236 tons of standard coal/10,000 RMB Yuan, 0.5992 tons of standard coal/10,000, 0.6235 tons of standard coal/10,000 RMB Yuan, 0.6146 tons of standard coal/10,000 RMB Yuan, and 0.6235 tons of standard coal/10,000 RMB Yuan, respectively. The total energy utilization is 7212.29 million tons of standard coal, and the energy utilization intensity is 0.6315 tons of standard coal/10,000 RMB Yuan.

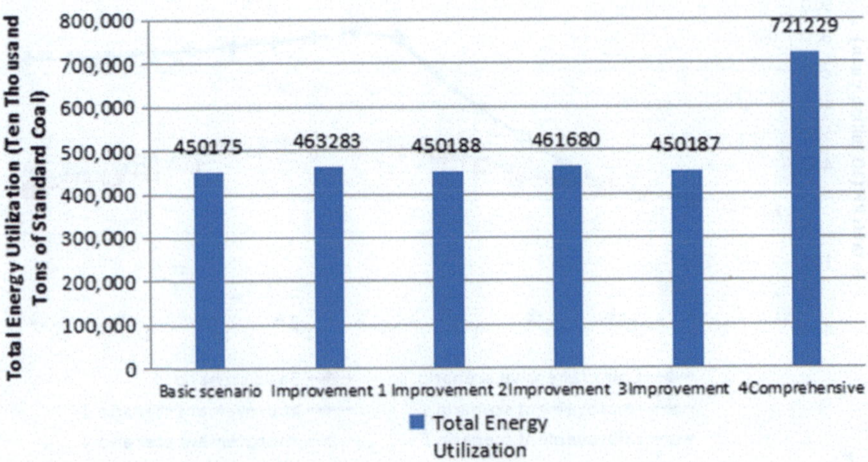

Fig. 4.3 Total energy utilization in Hebei province during the simulated period

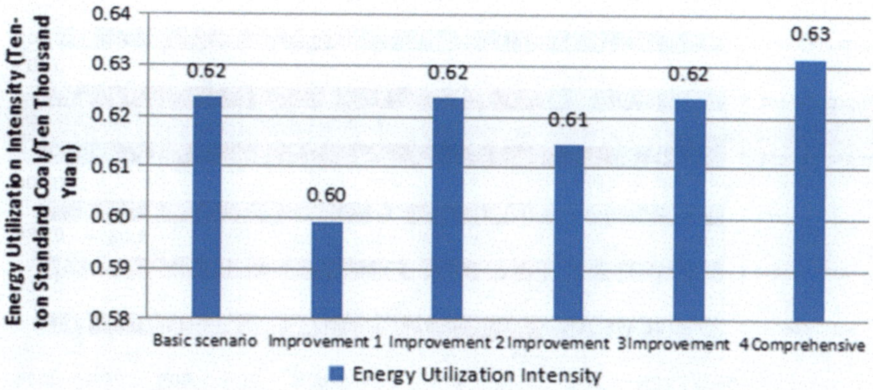

Fig. 4.4 Energy utilization intensity of Hebei province in the simulated period

According to the energy planning of Hebei province, the annual growth rate of total energy supply should not exceed 2.2% of the total energy utilization in the previous year, the annual supply of clean energy should not exceed 10% of the total energy, and the total energy supply should be greater than or equal to the total energy demand. The total energy demand is determined by economic development. The simulation results show that the energy utilization intensity of each scenario is not significantly different, ranging from 0.60 to 0.63. Although the total amount of energy use varies greatly, the efficiency of energy use varies little.

2. Analysis of energy utilization structure

The energy utilization structure of Hebei province in 2012 is shown in Fig. 4.5. In 2012, the total demand of coal, oil, natural gas, traditional electricity, and clean energy accounted for 88.70, 7.69, 1.94, 1.57, and 0.11% of the total energy use.

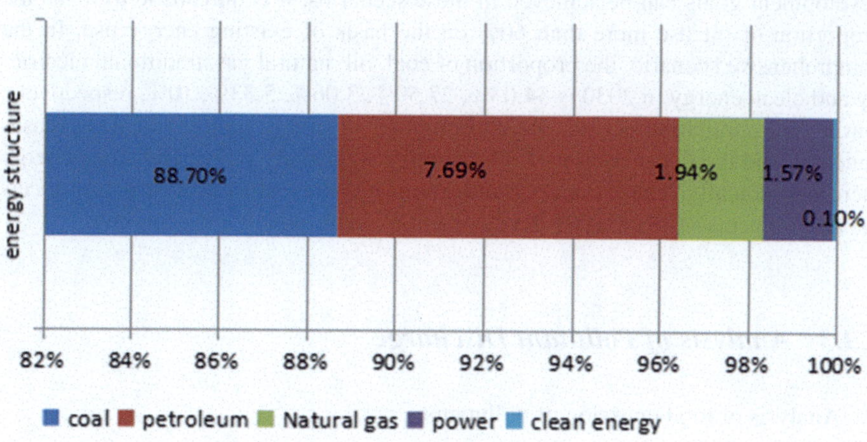

Fig. 4.5 Energy utilization structure of Hebei province in 2012

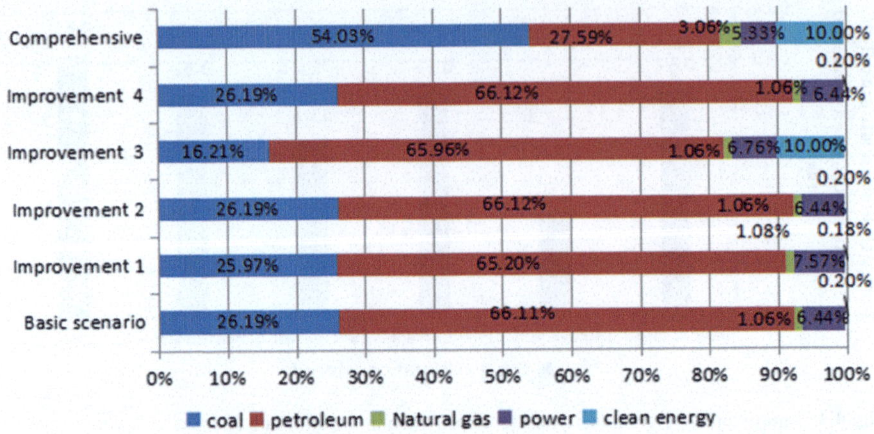

Fig. 4.6 Energy utilization structure in Hebei province in 2030

The energy utilization in Hebei province is dominated by coal, which causes serious pollution, and the utilization of clean energy is too low, which is harmful to the sustainable development of environment and economy.

Figure 4.6 shows the energy utilization structure of each scenario in Hebei province in 2030. The proportion of coal, oil, natural gas, traditional electricity and clean energy in the basic scenario, improved scenarios1, 2, 3, and 4 is 26.19%, 66.11%, 1.06%, 6.44%, 0.20%; 25.97%, 65.20%, 1.08%, 7.57%, 0.18%; 26.19%, 66.12%, 1.06%, 6.44% and 0.20%; 16.21%, 65.96%, 1.06%, 6.76%, and 10%; 26.19%, 66.12%, 1.06%, 6.44%, and 0.20%, respectively.

The results show that in order to meet the requirements of environmental restrictions, the consumption of coal and petroleum in basic scenarios and improved scenarios 1, 2, 3, and 4 is greatly reduced. Although economic, social and environmental development goals can be achieved in these scenarios, it is unrealistic to make the proportion of oil use more than 60% on the basis of existing energy use. In the comprehensive scenario, the proportion of coal, oil, natural gas, traditional electricity and clean energy in 2030 is 54.03%, 27.59%, 3.06%, 5.33%, 10%, respectively. During the simulation period, the consumption of highly polluted coal decreases annually, and the consumption of oil, natural gas, primary power and clean energy increases annually. The rational optimization of energy utilization structure meets the requirements of sustainable development of environment and economy.

4.2.3 Analysis of Pollutant Discharge

1. Analysis of total emission of pollutants

Environmental pollution in Hebei province is caused by industrial production, final consumption, motor vehicle exhaust, agricultural activities and other factors. The model establishes that the total $PM_{2.5}$ emission of Hebei province in 2030 is 30% less than that in 2016, and decreases in the same proportion every year. The following is the pollutant emission in each scenario during the simulation period.

The total emissions of SO_2, NOx, CO, VOCs, PM_{10}, $PM_{2.5}$ and GHG in the simulation period in the basic scenario, improvement scenarios 1, 2, 3, and 4 and comprehensive scenario are 27.6 million tons, 17.6 million tons, 146.9 million tons, 26.4 million tons, 20.5 million tons, 16.8 million tons and 14.1 billion tons; 27.2 million tons, 17.2 million tons, 137.7 million tons, 25.4 million tons, 19.74 million tons, 16.8 million tons and 14.1 billion tons; 27.6 million tons, 17.6 million tons, 146.5 million tons, 26.0 million tons, 20.5 million tons, 16.8 million tons and 14.0 billion tons; 27.8 million tons, 17.7 million tons, 147 million tons, 26.6 million tons, 20.7 million tons, 16.8 million tons and 13.9 billion tons; 27.6 million tons, 17.6 million tons, 146.5 million tons, 26.0 million tons, 20.5 million tons, 16.8 million tons and 14.0 billion tons; 33.6 million tons, 19.84 million tons, 142.5 million tons, 29.4 million tons, 27.0 million tons, 16.8 million tons and 19.5 billion tons, respectively (see Figs. 4.7, 4.8, 4.9, 4.10, 4.11, 4.12 and 4.13).

The simulation results show that the goal of reducing $PM_{2.5}$ emission by 30 percent in 2030 compared with 2016 could be achieved with the adoption of environmental policies in all six scenarios. However, due to the different emission coefficients of pollutants, economic developments, and various policies, the emissions of SO_2, NOx, PM_{10} and VOCs in the comprehensive scenario are higher than those in other scenarios.

2. Analysis of emission intensity of pollutants

Emission intensity of pollutants refers to the emission amount of pollutants per unit GRP, and the unit is generally per ton/RMB 10,000 Yuan GRP. It is a critical indicator to measure economic and social development and environmental improvement. Figures 4.14, 4.15, 4.16, 4.17, 4.18, 4.19 and 4.20 are the emission intensity of SO_2,

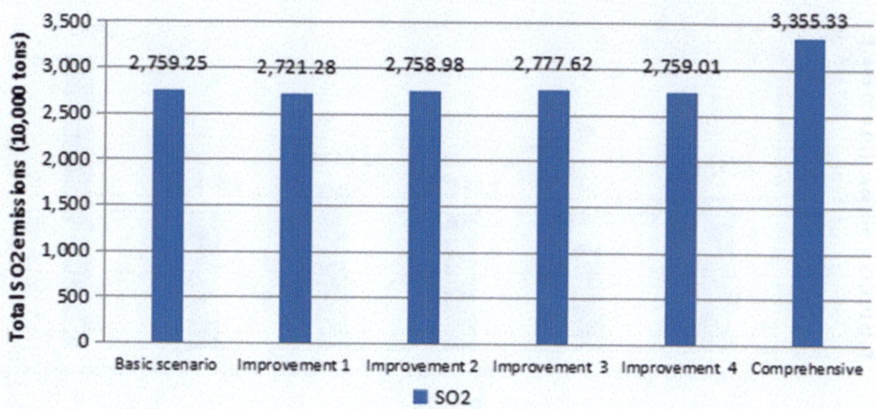

Fig. 4.7 Total SO_2 emission in Hebei province during the simulated period

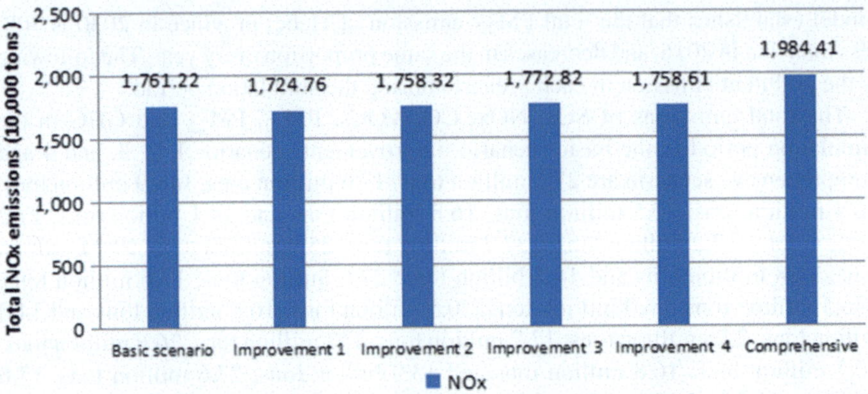

Fig. 4.8 Total NOx emission in Hebei province during the simulated period

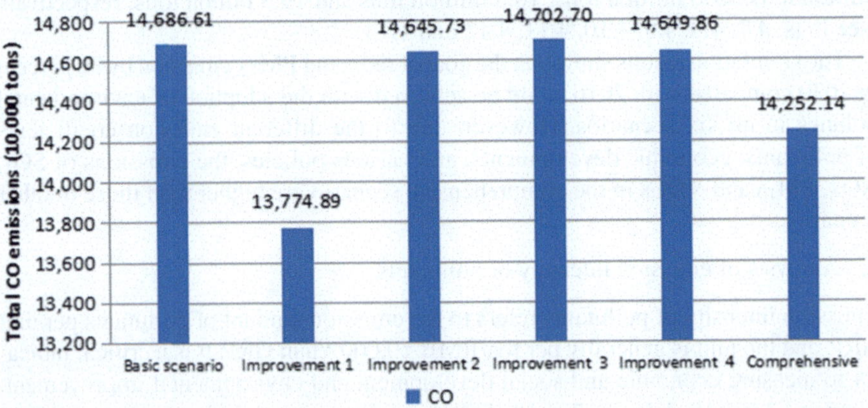

Fig. 4.9 Total CO Emission in Hebei province during the simulated period

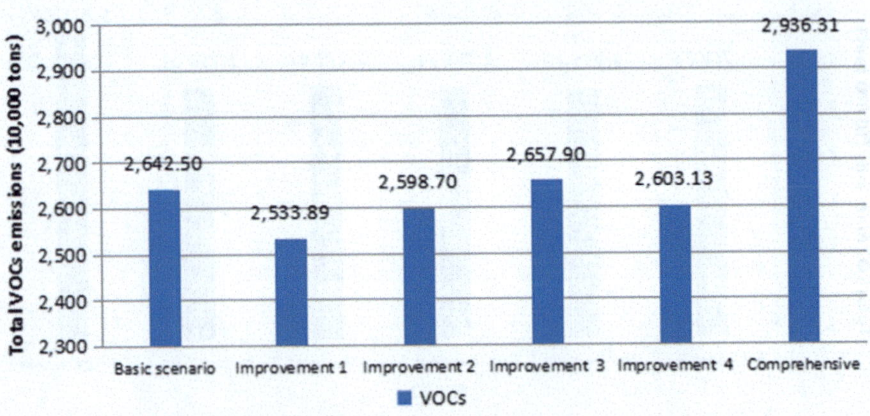

Fig. 4.10 Total VOCs emission in Hebei province during the simulated period

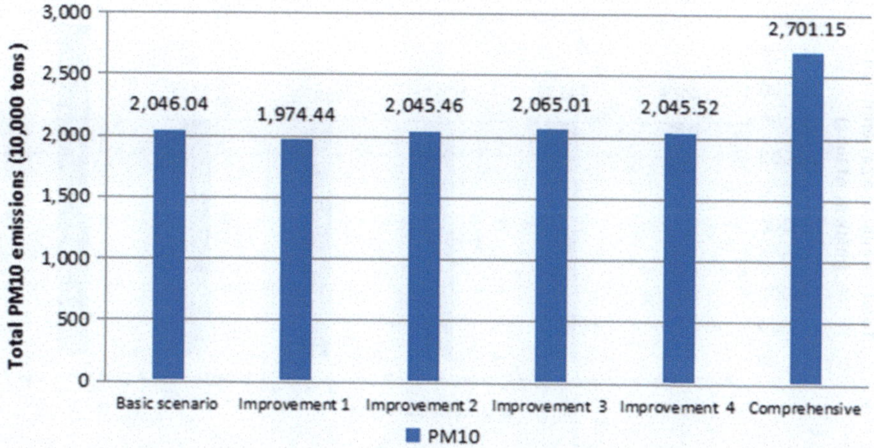

Fig. 4.11 Total PM_{10} emissions in Hebei province during the simulated period

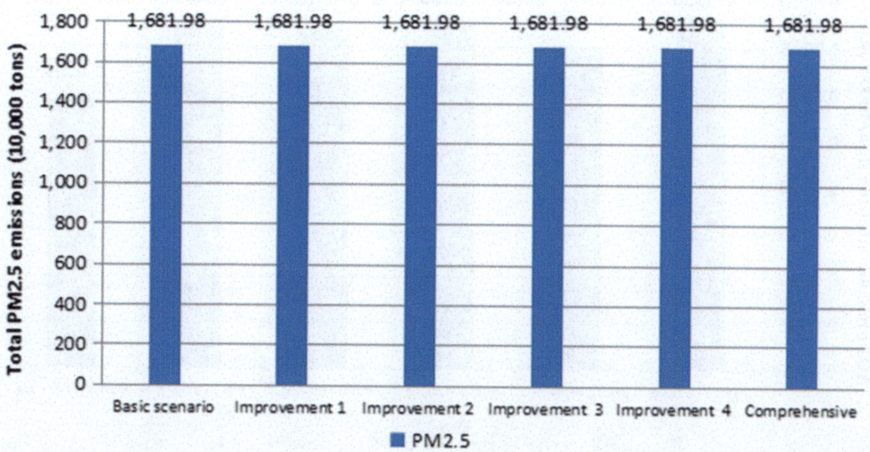

Fig. 4.12 Total $PM_{2.5}$ emission in Hebei province during the simulation period

NOx, PM_{10}, VOCs, $PM_{2.5}$, CO and GHG during the simulation period of Hebei province.

During the simulation period, the average emission intensity of SO_2 in basic scenario, improved scenarios1–4, and comprehensive scenario is 0.0038 tons/RMB 10,000 Yuan GRP, 0.0035 tons/RMB 10,000 Yuan GRP, 0.0038 tons/RMB 10,000 Yuan GRP, 0.0037 tons/RMB 10,000 Yuan GRP, 0.0038 tons/RMB 10,000 Yuan GRP and 0.0029 tons/RMB 10,000 Yuan GRP, respectively.

The average emission intensity of NO during the simulation period in the basic scenario, improved scenarios1–4, and comprehensive scenario is 0.002440 tons/RMB 10,000 Yuan GRP, 0.002231 tons/RMB 10,000 Yuan GRP, 0.002435 tons/RMB

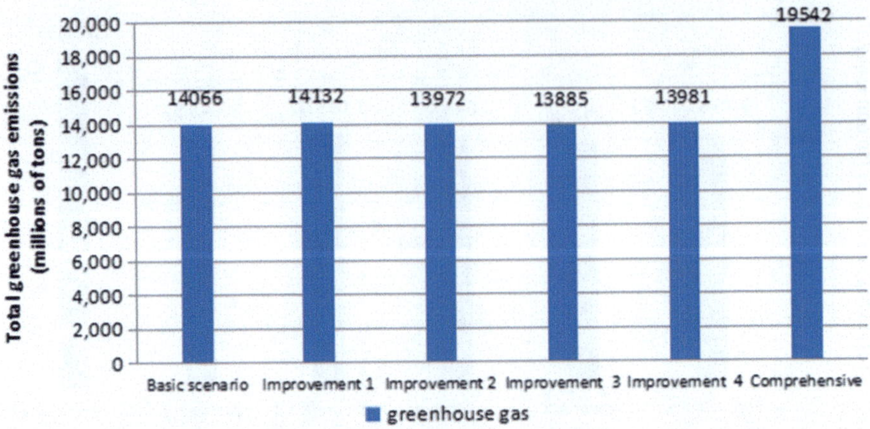

Fig. 4.13 Total GHG emissions in Hebei province during the simulated period

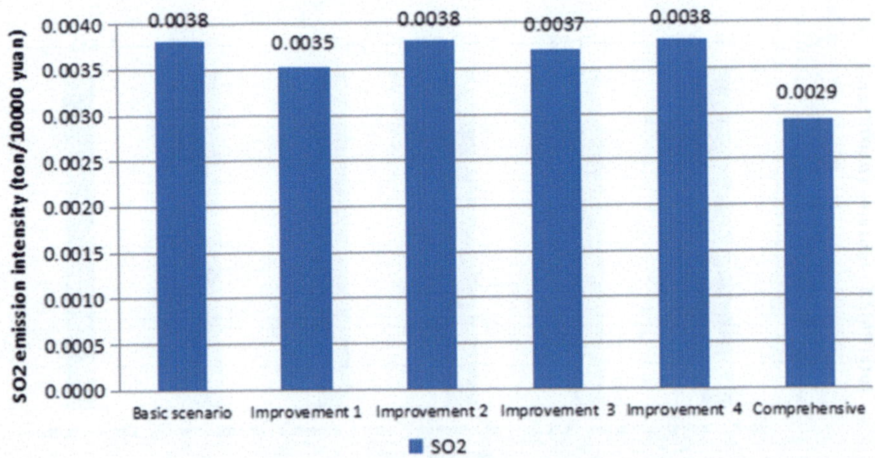

Fig. 4.14 SO_2 emission intensity in Hebei province during the simulated period

10,000 Yuan GRP, 0.002326/RMB 10,000 Yuan GRP, 0.002436 tons/RMB 10,000 Yuan GRP and 0.001738 tons/RMB 10,000 Yuan GRP, respectively.

The average emission intensity of simulated PM_{10} in basic scenarios1–4, and comprehensive scenario is 0.0028 tons/RMB 10,000 Yuan GRP, 0.0026 tons/RMB 10,000 Yuan GRP, 0.0028 tons/RMB 10,000 Yuan GRP, 0.0027 tons/RMB 10,000 Yuan GRP, 0.0028 tons/RMB 10,000 Yuan GRP and 0.0024 tons/RMB 10,000 Yuan GRP, respectively.

The average emission intensity of simulated CO in basic scenario, improved scenarios 1, 2, 3, and 4, and integrated scenario is 0.0203 tons/RMB 10,000 Yuan GRP, 0.0178 tons/RMB 10,000 Yuan GRP, 0.0203 tons/RMB 10,000 Yuan GRP,

Fig. 4.15 NO_X emission intensity in Hebei province during the simulated period

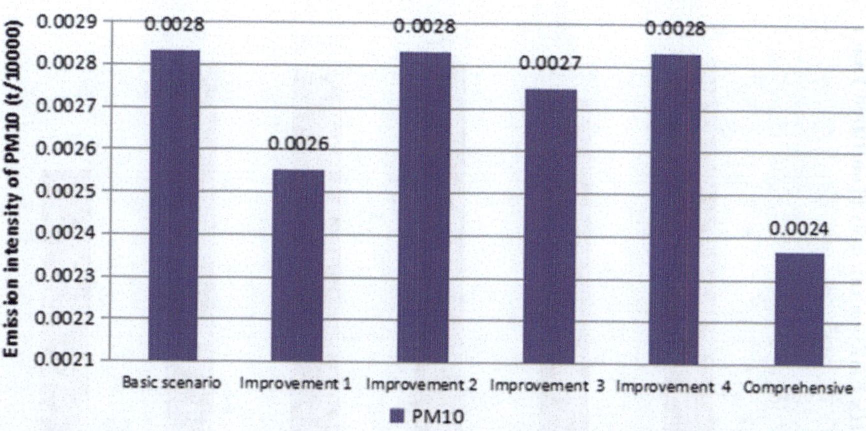

Fig. 4.16 Emission intensity of PM_{10} in Hebei province during the simulated period

0.0196 tons/RMB 10,000 Yuan GRP, 0.0203 tons/RMB 10,000 Yuan GRP and 0.0125 tons/RMB 10,000 Yuan GRP, respectively.

The average emission intensity of simulated VOCs in basic scenario, improved scenarios1–4 and comprehensive scenario is 0.0037 tons/RMB 10,000 Yuan GRP, 0.0033 tons/RMB 10,000 Yuan GRP, 0.0036 tons/RMB 10,000 Yuan GRP, 0.0035 tons/RMB 10,000 Yuan GRP, 0.0036 tons/RMB 10,000 Yuan GRP and 0.0026 tons/RMB 10,000 Yuan GRP, respectively.

The average emission intensity of simulated $PM_{2.5}$ in basic scenario, improved scenarios 1, 2, 3, and 4, and comprehensive scenario is 0.002330 tons/RMB 10,000 Yuan GRP, 0.002176 tons/RMB 10,000 Yuan GRP, 0.002330 tons/RMB 10,000 Yuan

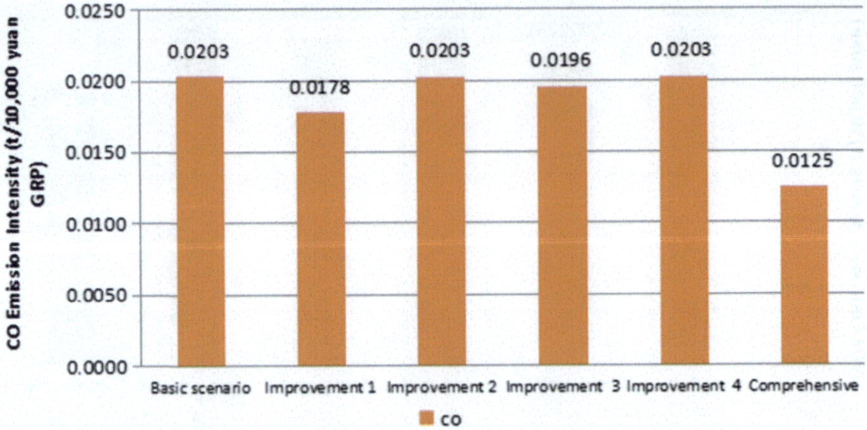

Fig. 4.17 CO emission intensity in Hebei province during the simulated period

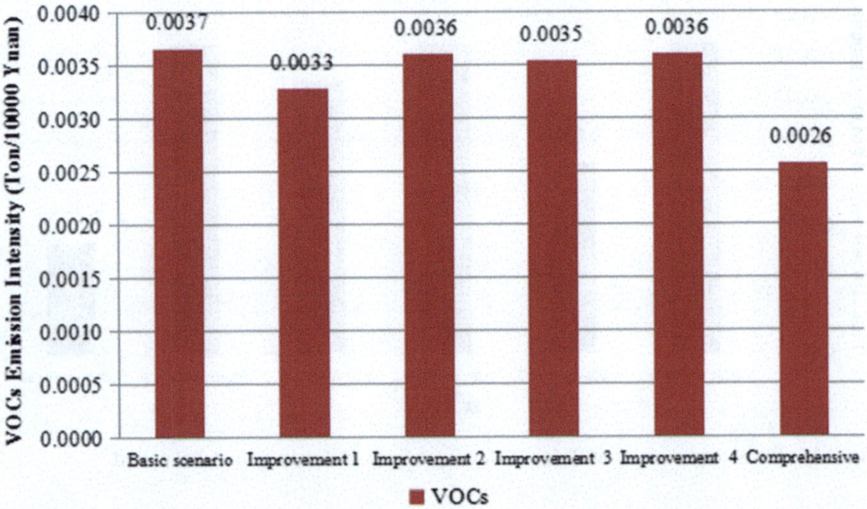

Fig. 4.18 VOCs emission intensity in Hebei province during the simulated period

GRP, 0.002239 tons/RMB 10,000 Yuan GRP, 0.00233,000 tons/RMB 10,000 Yuan GRP and 0.001473/RMB 10,000 Yuan GRP, respectively.

The average emission intensity of simulated GHG in basic scenario, improved scenarios 1, 2, 3, and 4, and integrated scenario is 19 483 tons/RMB 10 000 Yuan GRP, 18 279 tons/RMB 10,000 Yuan GRP, 19 352 tons/RMB 10,000 Yuan GRP, 18 483 tons/RMB 10,000 Yuan GRP, 19 366 tons/RMB 10,000 Yuan GRP and 17 110 tons/RMB 10,000 Yuan GRP, respectively.

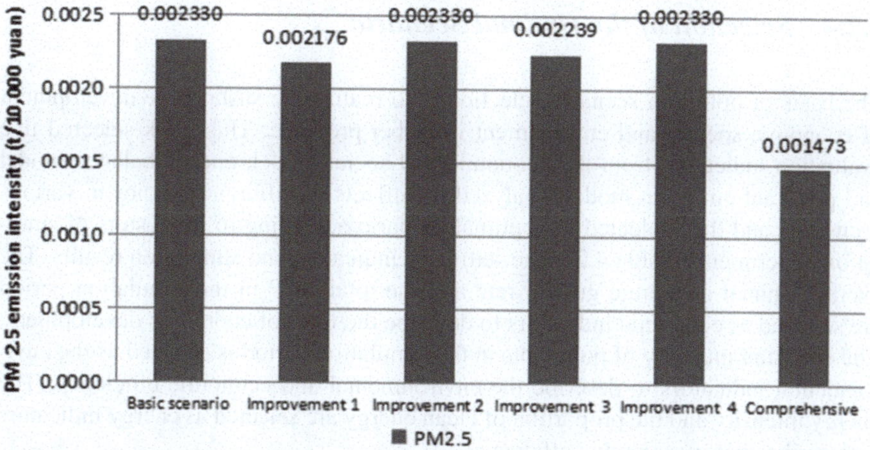

Fig. 4.19 Emission intensity of $PM_{2.5}$ in Hebei province during the simulated period

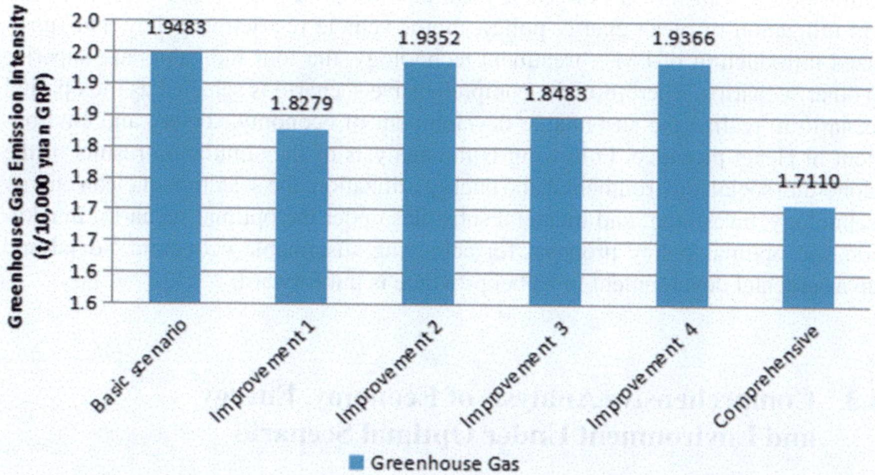

Fig. 4.20 GHG emission intensity in Hebei province during the simulated period

The emission intensity of six pollutants in the comprehensive scenario is lower than that in other scenarios. The simulation results show that the environmental and economic efficiency of the single environmental improvement policy is lower than that of the comprehensive environmental improvement policy. The comprehensive environmental improvement policy adopted in the comprehensive scenario is more efficient to promoting economic growth while achieving emission reduction targets.

4.2.4 Selection of the Optimal Scenario

The basis of optimum scenario selection is to realize the sustainable development of economy, society and environment in Hebei province. This book selected five evaluation indicators from the economic and social model, energy balance model and pollutant emission model, analyzed the effects of different policies in various scenarios, and then selected the optimal scenario according to the results of simulation experiments. Table 4.2 is the setting of indicators and simulation results. The average annual economic growth rate and the total GRP in the simulation period are selected as economic indicators to describe the level of economic development; The emission intensity of pollutants in the simulation period is selected as the environmental indicators to describe the environmental and economic efficiency; The energy intensity and the proportion of clean energy are selected as energy indicators to describe energy economic efficiency.

According to the simulation results, in the comprehensive scenario, by introducing the policy combination of subsidized industrial restructuring, soil and water conservation policy, subsidized new energy automobile policy, subsidized development and utilization of clean energy policy, motor vehicle restriction policy and subsidized introduction of $PM_{2.5}$ treatment technology, the four indicators are superior to other scenarios. Therefore, the comprehensive scenario is selected as the optimal scenario to realize the sustainable development of economy, society and environment in Hebei province. Following is the analysis of the simulation results of the economic, social, environmental and energy utilization, the selection of clean energy technology, investment and financial subsidies under the optimal scenario. In addition, the optimal policy proposal for achieving sustainable economic, social and environmental development in Hebei province is put forward.

4.3 Comprehensive Analysis of Economy, Energy and Environment Under Optimal Scenario

4.3.1 Analysis of Economic Growth

1. Analysis of Total Economic Quantity and Economic Development Trend

The change of GRP of Hebei from 2012 to 2030 is shown in Fig. 4.21. The GRP of Hebei province increases first and then decreases. The economic growth is faster especially after the adoption of the soil and water conservation policy, the subsidy for new energy vehicles, the subsidy for the development and utilization of clean energy policies, the subsidy for introducing $PM_{2.5}$ treatment technology policies, and the policy of motor vehicle restrictions in 2017–2021. After 2022, due to the bottleneck of policy and technology emission reduction effect, the pressure of emission reduction is greater, and the development of economy shows a downward trend. Overall, the

Table 4.2 Evaluation indicators for optimal scenario selection

Index	Indicator description		Basic scenario	Improvement scenario 1	Improvement scenario 2	Improvement scenario 3	Improvement scenario 4	Comprehensive scenario
Economic indicators	Total GRP (trillion)		72.19	77.31	72.20	75.12	72.20	**114.21**
	Average annual economic growth (%)		1.38	2.56	1.38	1.87	1.38	**5.87**
Energy indicators	Energy utilization intensity (Ten-ton Standard Coal/RBM 10,000 Yuan GRP)		0.62	0.60	0.62	0.61	0.62	0.63
	Clean energy share by 2030 (%)		0.14	0.13	0.14	5.19	0.14	**10**
Environmental indicators	Emission Intensity of pollutants (t/RMB 10,000 Yuan GRP)	SO_2	0.0038	0.0035	0.0038	0.0037	0.0038	**0.0029**
		NO_x	0.0024	0.0022	0.0024	0.0024	0.0024	**0.0017**
		PM_{10}	0.0028	0.0026	0.0028	0.0027	0.0028	**0.0024**
		CO	0.00203	0.00178	0.00203	0.00196	0.00203	**0.00125**
		VOCs	0.0037	0.0033	0.0036	0.0035	0.0036	**0.0026**
		$PM_{2.5}$	0.0023	0.0022	0.0023	0.0022	0.0023	**0.0015**

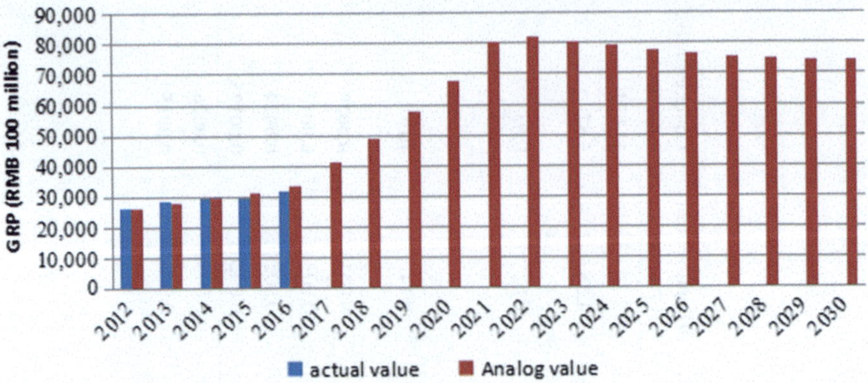

Fig. 4.21 Changes of GRP in Hebei province from 2012 to 2030

GRP of Hebei province increases at an average annual growth rate of 5.87% from 2.66 trillion RMB Yuan in 2012 to 7.42 trillion RMB Yuan in 2030, with a total increase of 179.14%.

In the meanwhile, it should be noted that the actual value of GRP in Hebei province in 2013 and 2014 is slightly higher than the simulation value. This is because the GRP in the model is calculated based on the year 2012, and the price changes were excluded. While the actual value of GRP in 2013 and 2014 is calculated on the basis of the previous year's price. The simulated value would be closer to the actual value if the price factor was excluded.

2. Analysis on the change of output value and structure of three industries

The structure of the three industries of Hebei province in 2012 and 2030 are shown in Fig. 4.22. With the development of economy and society, efficient use of energy and environmental protection, the structure of the three industries in Hebei province

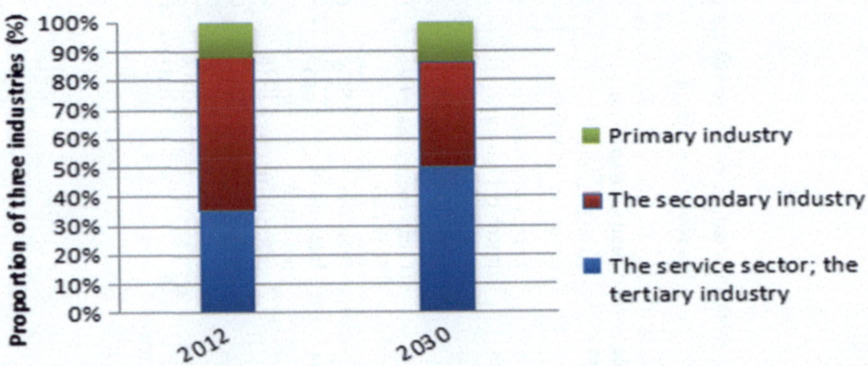

Fig. 4.22 Proportion of three industries in Hebei province in 2012 and 2030

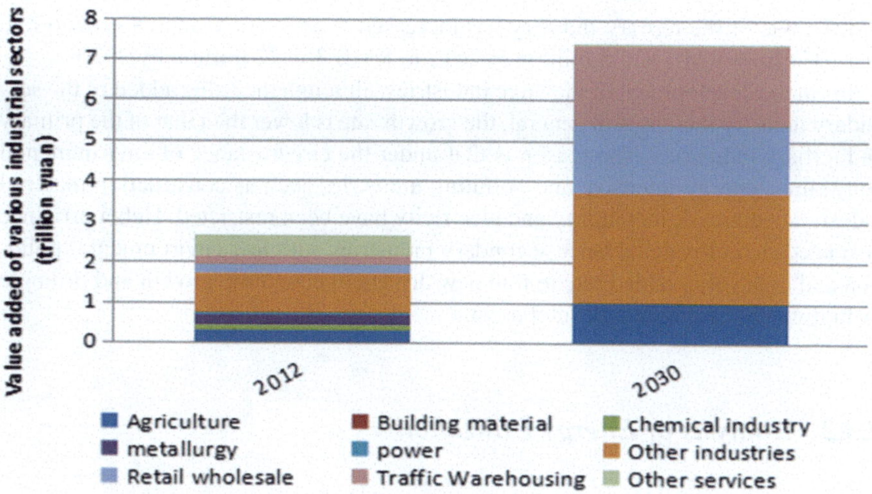

Fig. 4.23 Added value of industrial departments of Hebei province in 2012–2030

have been optimized annually. From 2012 to 2030, the proportion of primary industry remains basically unchanged, the proportion of secondary industry gradually decreases, and the proportion of tertiary industry gradually increases. The proportion of the three industries is optimized from 12:53:35 in 2012 to 13:37:50 in 2030.

The added value of each industry of Hebei province in 2012 and 2030 is shown in Fig. 4.23. From the perspective of the development of various sectors within the industry, the output value of the primary industry has steadily increased due to the low pollutant emission coefficient of the industry, and the lower total amount of energy utilization, energy intensity and pollutant emission compared with other sectors. Therefore, the primary industry has been developed under the constraints of energy utilization and environmental protection. The output value of the primary industry has increased at an average annual growth rate of 6.55% from RMB 31.87 million in 2012 to 99.81 RMB million by 2030.

Under the constraints of resources and environmental carrying capacity, the output value of construction material industry, chemical industry, metallurgy and electric power with high pollution and energy consumption in the secondary industry has decreased significantly annually. The output value of other industries with less energy consumption and pollution increases annually. With the implementation of the various policies, the internal structure of the secondary industry has also been optimized annually. The added value of the second industry has increased at an average annual growth rate of 3.76% from RMB 140.04 million in 2012 to 272.29 RMB million by 2030.

Retail and wholesale, transportation and storage, low energy demand coefficient and low emission intensity of pollutants in the tertiary industry have been well developed under the constraints of resource utilization and environmental protection. The added value of the tertiary industry has increased at an average annual growth rate of 7.91% from RMB 93.85 million in 2012 to RMB 369.72 million by 2030.

From the development of the three industries, although the value added of the secondary industry increases in general, the growth rate is lower than that of the primary and tertiary industries. The reason is that under the circumstance of environmental constraints, energy-intensive and polluting industries such as construction material industry, chemicals, metallurgy and electricity have been restricted. Hebei province also needs to cultivate advance secondary industries with low environmental pollution and high value-added rate to find new drivers of economic growth and promote its high-quality economic development.

4.3.2 Analysis of Energy Utilization

1. Trends in energy utilization structure

The changes in total energy use and energy structure in Hebei province from 2012 to 2030 are shown in Fig. 4.24 and Table 4.3, respectively. The model assumes that the increase in energy supply in Hebei province from 2012 to 2030 shall not exceed 2.2% of the previous year. Because of the serious pollution of coal, the model assumes that the increase of clean energy would replace the amount of coal without affecting other energy sources.

Due to the implementation of various environmental policies, Hebei's energy structure has been gradually optimized after 2017. The proportion of coal decreases annually, while the proportion of other energy sources increases annually. From 2017 to 2030, the average annual use of clean energy accounts for 10% of the total energy use. In 2016, coal, oil, natural gas, traditional electricity and clean energy accounts

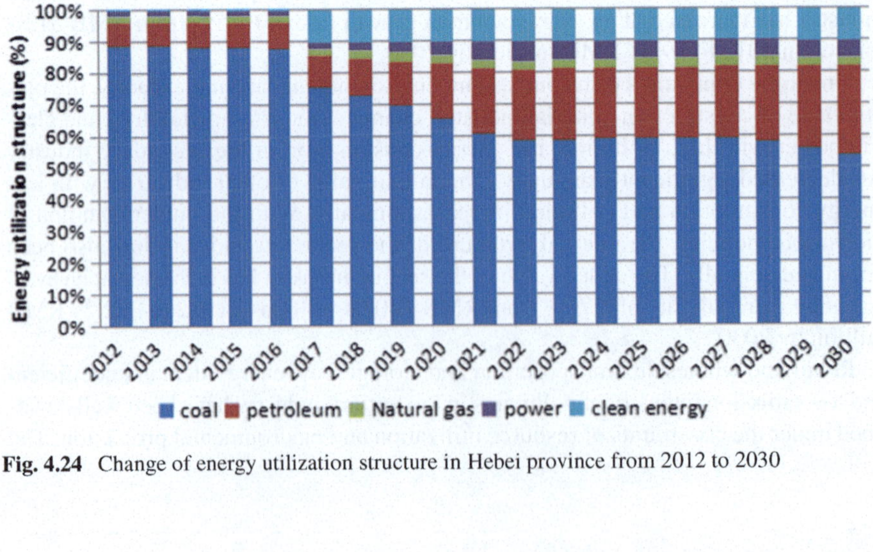

Fig. 4.24 Change of energy utilization structure in Hebei province from 2012 to 2030

Table 4.3 Energy use in Hebei province, 2012–2030 (Unit: 10,000 tons of standard coal)

Year (year)	Coal	Petroleum	Natural gas	Traditional electricity	Clean energy	Total
2012	26865.04	2325.196	585.665	474.3324	32.43992	30250.23
2013	28343.98	2469.774	618.8236	506.4847	32.43992	31971.5
2014	28852.54	2600.939	648.1236	540.8372	32.43992	32674.88
2015	29363.48	2740.902	679.0582	577.8485	32.43992	33393.73
2016	29876.15	2890.333	711.725	617.7396	32.43992	34128.39
2017	26239.77	3508.985	842.0045	800.5291	3487.921	34879.21
2018	25894.36	4232.897	930.6135	1024.032	3564.656	35646.56
2019	25344.64	5140.286	985.4265	1317.354	3643.078	36430.78
2020	24442.53	6336.134	1009.658	1720.704	3723.226	37232.25
2021	22985.27	7966.235	1003.305	2291.422	3805.137	38051.37
2022	22757.94	8701.271	1044.155	2496.284	3888.85	38888.5
2023	23347.51	8862.488	1104.291	2455.352	3974.404	39744.05
2024	23896.04	9078.465	1160.138	2421.929	4061.841	40618.41
2025	24475.16	9298.388	1214.693	2372.572	4151.202	41512.02
2026	25042.75	9555.558	1268.197	2318.242	4240.535	42425.28
2027	25577.69	9876.69	1320.849	2266.834	4316.574	43358.64
2028	25332.25	10357.6	1343.031	2259.701	4316.574	43609.16
2029	24270.3	11022.36	1337.409	2302.08	4316.574	43248.72
2030	23320.42	11910.02	1320.126	2298.608	4316.574	43165.75

for 87.54%, 8.47%, 2.09%, 1.81% and 0.10% of the total energy use, respectively, and that in 2030 is 54.03%, 27.59%, 3.06%, 5.33% and 10%, respectively. Due to the dramatic reduction of the total amount of high-pollution coal utilization and the optimization of other energy utilization structure, the utilization structure of energy structure in Hebei province is optimized.

2. Trend of energy utilization intensity

The trend of energy utilization intensity in Hebei province from 2012 to 2030 is shown in Fig. 4.25. From the simulation results, it can be seen that the energy utilization intensity of Hebei province gradually decreases from 2012 to 2022, increases slightly from 2023 to 2030, and maintains at 0.5-0.6 tons of standard coal/RMB 10,000 Yuan GRP. In 2012, the energy utilization intensity was 1.14 tons of standard coal/RMB 10,000 Yuan GRP, which is reduced by 49% to 0.58 tons of standard coal/RMB 10,000 Yuan GRP in 2030, and the energy utilization efficiency is greatly improved. In the meanwhile, it proves that the comprehensive policy combination of energy utilization and environmental improvement can effectively promote economic development.

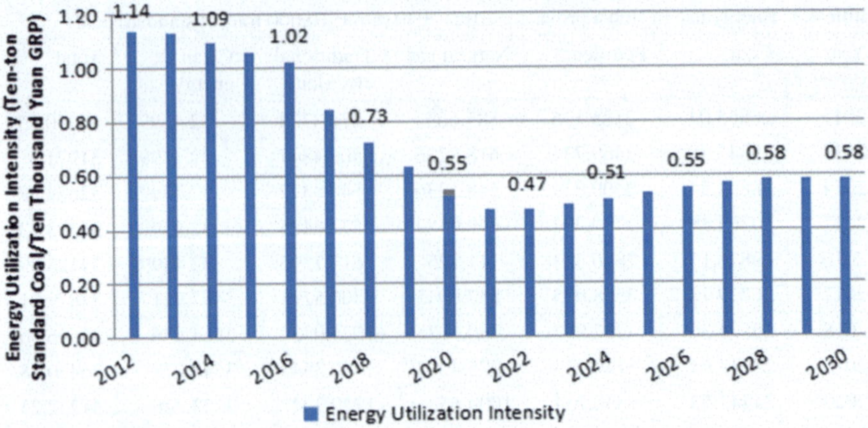

Fig. 4.25 Trends of energy utilization intensity in Hebei province from 2012 to 2030

4.3.3 Analysis of Pollutant Emission

1. Analysis of total emission of pollutants

Emission of pollutants in Hebei province from 2012 to 2030 is shown in Figs. 4.26, 4.27, 4.28, 4.29, 4.30, 4.31 and 4.32. $PM_{2.5}$ emission in 2030 is set to decrease by 30% compared with 2016. From these simulation results, we can conclude that

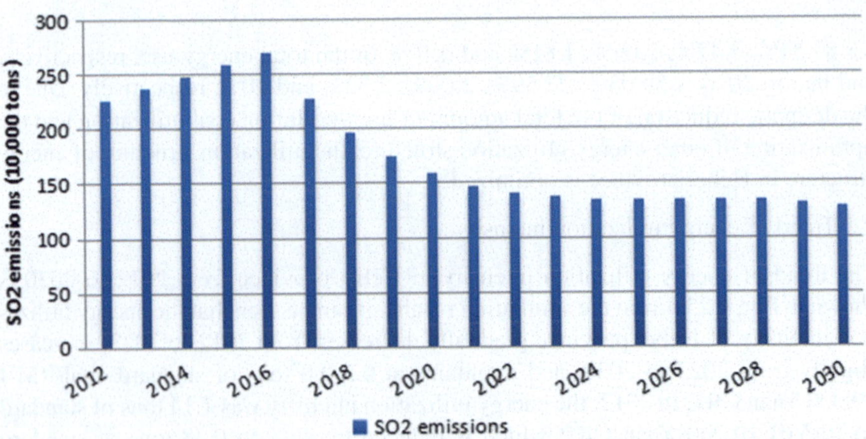

Fig. 4.26 Change trend of SO_2 emission in Hebei province from 2012 to 2030

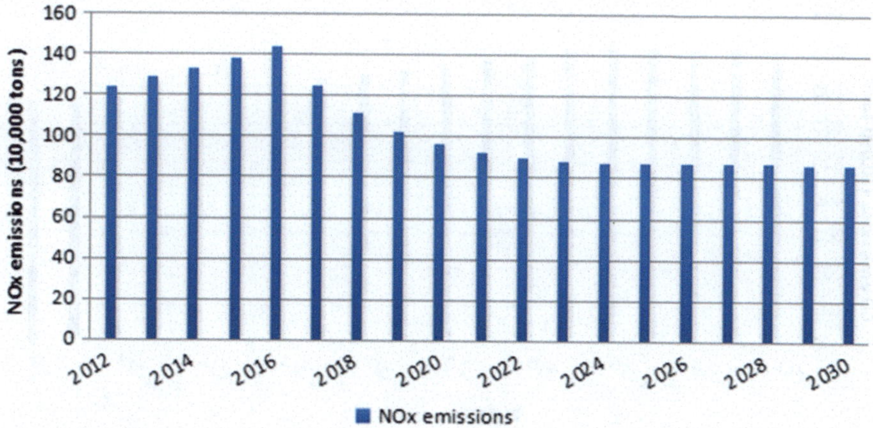

Fig. 4.27 Change trend of NOx Emission in Hebei province from 2012 to 2030

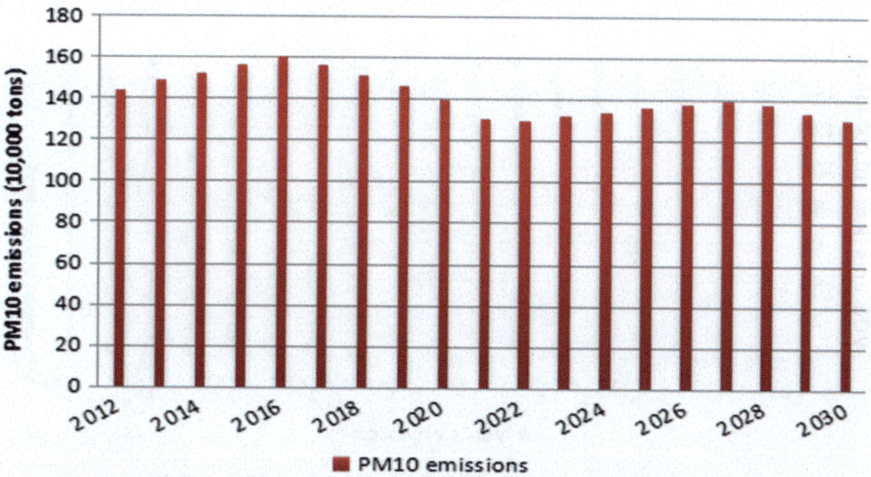

Fig. 4.28 Change trend of PM_{10} emission in Hebei province from 2012 to 2030

all scenarios achieve the goal of the model. However, due to the different emission coefficients of pollutants, the effects of various policies on pollutant emission reduction are also distinct, leading to some differences in the total amount of pollutant emissions. During the simulation period, the emissions of SO_2, NO, PM_{10}, CO, $PM_{2.5}$ and GHG increases annually from 2012 to 2016. Due to the implementation of haze control policy, the emission of six pollutants decreases significantly annually after 2017. Compared with 2016, the emissions of SO_2, NOx, PM_{10}, CO, VOCs, $PM_{2.5}$ and GHG decreased by 52%, 40%, 19%, 18%, 6%, 32%, and -3.3% in 2017, respectively. It shows that the implementation of environmental policy has

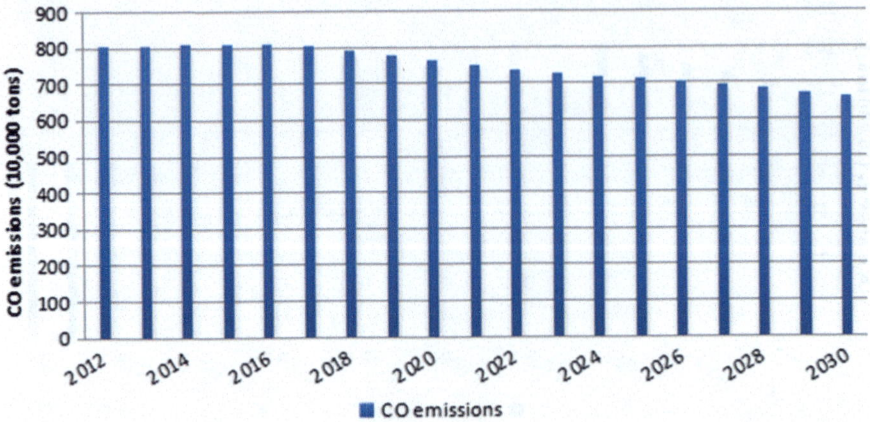

Fig. 4.29 Trends of CO emission in Hebei province from 2012 to 2030

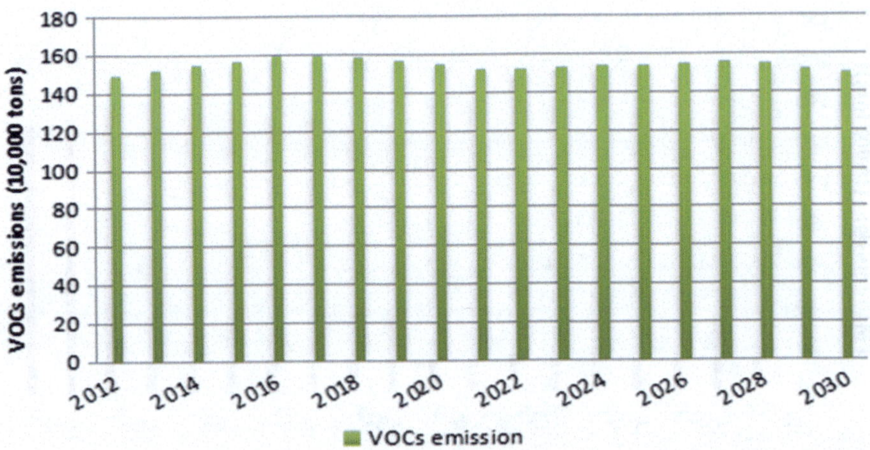

Fig. 4.30 Change trend of VOCs emission in Hebei province from 2012 to 2030

marvelous effect on environmental improvement. Because the emission coefficients of different pollutants are distinct, the reductions also vary.

2. Analysis of emission intensity of pollutants

The economy grows rapidly from 2017 to 2021, while the emission of pollutants decreases, therefore, the emission intensity of pollutants decreases rapidly, and the effect of haze control policy is excellent. The economic growth has slowed down since 2022, and the intensity of pollutant emission changes smoothly, and bottlenecks in the effectiveness of smog control policies emerges. Compared with 2016, the emission of SO_2, NOx, PM_{10}, CO, VOCs, $PM_{2.5}$ and GHG intensity decreases by 52%, 40%,

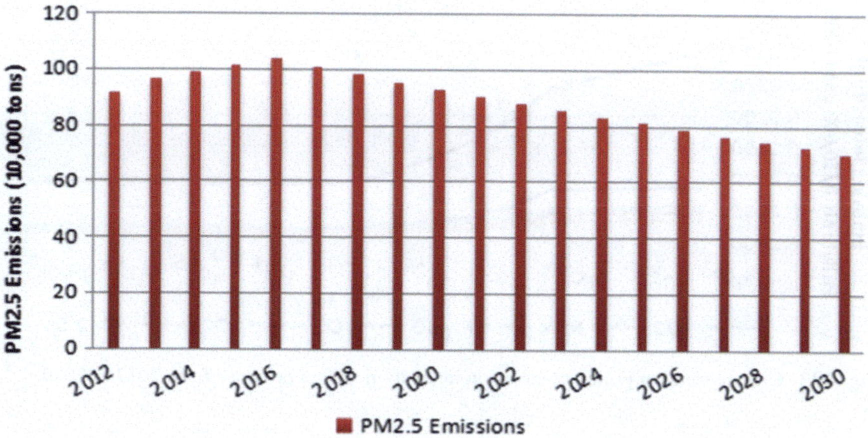

Fig. 4.31 Change trend of $PM_{2.5}$ emission in Hebei province from 2012 to 2030

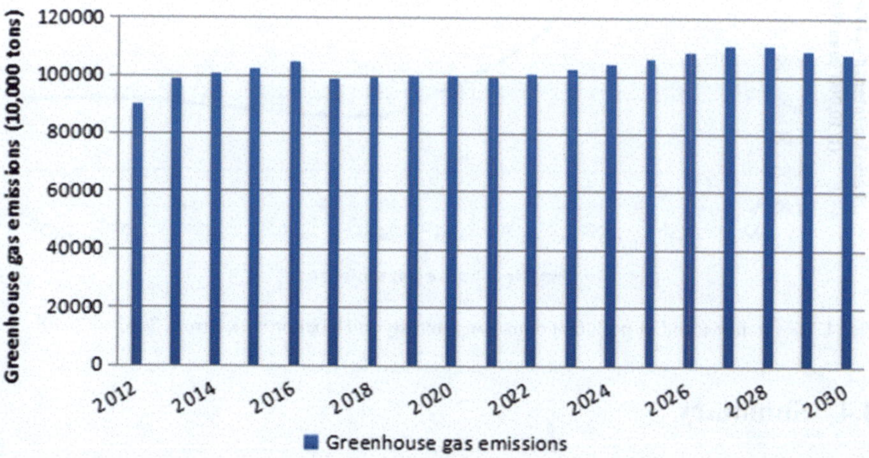

Fig. 4.32 Trends of GHG emission in Hebei province from 2012 to 2030

19%, 18%, 6%, 32% and -3.3%, respectively in 2030. The emission intensity of pollutants in Hebei province during the simulation period is shown in Figs. 4.33 and 4.34.

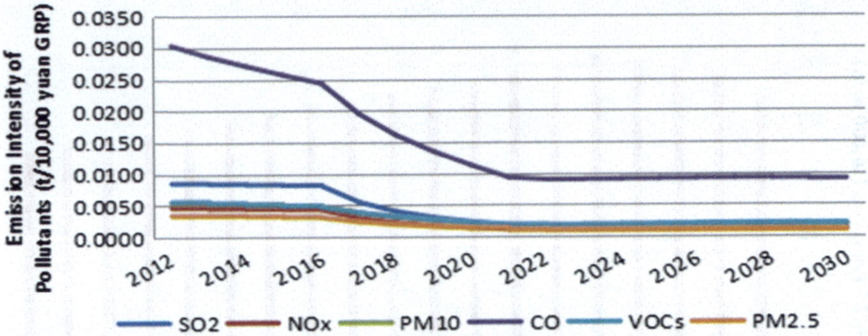

Fig. 4.33 Change trend of pollutant emission intensity in Hebei province from 2012 to 2030

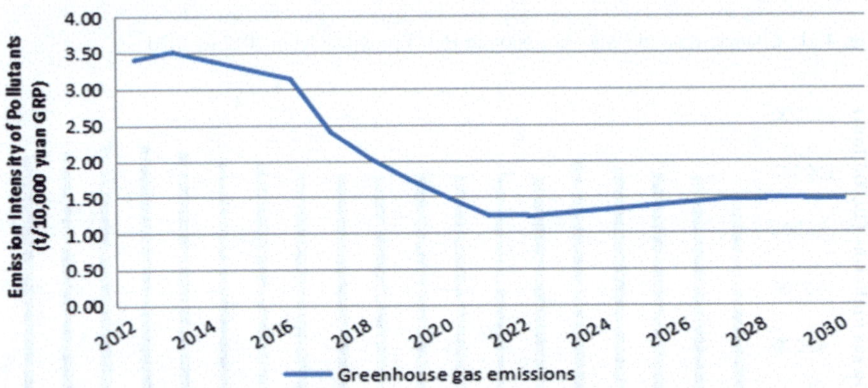

Fig. 4.34 Change trend of pollutant emission intensity in Hebei province from 2012 to 2030

4.4 Summary

This chapter compares and analyses five indicators of Hebei's economic system, energy system and environmental system under different scenarios, and selects the comprehensive scenario as the optimal scenario for realizing the sustainable development of economy, society, energy and environment of Hebei.

In the optimal scenario, the trend of economic changes in Hebei province shows a situation of first rising and then falling. By 2030, the total GRP of Hebei province will be 7.42 trillion RMB Yuan, and the average economic growth rate is 5.87% from 2012 to 2030. The economic structure has been optimized, and the proportion of three industries could reach 13:37:50 in 2030.

The structure of energy supply and demand is optimized, and the intensity of energy consumption is reduced. In 2030, coal, oil, natural gas, traditional electricity and clean energy accounts for 54.03%, 27.59%, 3.06%, 5.33% and 10% of the total energy utilization, respectively. The energy structure of Hebei province has been optimized due to the dramatic reduction of the total amount of coal utilization with high pollution and the dramatic increase of clean energy and the optimization of the structure of other energy utilization. The energy utilization intensity in 2030 is 0.58 tons of standard coal/RMB 10,000 Yuan GRP, which is 49% lower than that in 2012, and the energy utilization efficiency is greatly improved.

Environmental and economic benefits have gradually improved, and $PM_{2.5}$ emission has been reduced annually. The implementation of environmental policies has a significant effect on environmental improvement. Compared with 2016, the emission of SO_2, NOx, PM_{10}, CO, VOCs, $PM_{2.5}$ and GHG intensity decreases by 52%, 40%, 19%, 18%, 6%, 32% and -3.3%, respectively in 2030.

From 2017 to 2021, the comprehensive haze control policy has achieved remarkable results, and the economy has developed rapidly. From 2022 to 2030, there is a bottleneck in the effectiveness of policies to reduce emissions. In order to achieve sustained and rapid economic growth under the constraints of environmental emission reduction targets, new policies and technologies need to be further introduced.

From 2017 to 2021, the comprehensive haze control policy has achieved remarkable results, and the economy has developed rapidly. From 2022 to 2030, there is a bottleneck in the effectiveness of policies to reduce emissions. In order to achieve sustained and rapid economic growth under the constraints of environmental emission reduction targets, new policies and technologies need to be further introduced.

Chapter 5
Research Conclusions and Policy Recommendations

5.1 Research Conclusions

5.1.1 Status Quo Research Conclusion

Along with the rapid social economic development and the intensification of industrialization and urbanization, the pressure of energy and environmental carrying capacity in Hebei province has further intensified. In Hebei province, the economic activity of human society is the main cause of fog and haze and the unreasonable structure of industry and energy utilization are the main factors restricting the sustainable development of economy and environment. In the process of haze control in BTH region, Hebei province plays an important role, and the policies of haze control should be focused on Hebei province. Hebei province has issued a series of policies and measures to control haze pollution, but the economic, social and environmental impact of these policies and measures remains undiscovered.

5.1.2 Model Research Conclusion

This book constructs a comprehensive evaluation model of dynamic optimization of haze control policy. The model includes one objective function, that is, the maximization of economy growth and three sub-models including the economic growth model, the pollutant emission model, and the energy balance model. The theoretical basis of the model is the input-output model, material balance, value balance, and energy balance. Six haze control policies are written into the model as exogenous variables. By comparing the economic, environmental and energy indicators in different situations, the comprehensive effects of various policy combinations on the economy, energy and environment of Hebei province are analyzed. According to the simulation results, the optimal path to realize the sustainable development of economy, society, energy and environment in Hebei province was found out.

G. Zhang et al., *The Effect Evaluation of Haze Governance Policies in Hebei Province–Based on I-O Model*, https://doi.org/10.1007/978-981-15-3757-8_5

5.1.3 *Simulation Results*

Based on the data of economy, society, energy and environment in Hebei province in 2012, this book conducts a simulation experiment. The simulation period is 19 years from 2012 to 2030. The results show that the comprehensive policy combination of industrial restructuring policy, soil and water conservation policy, subsiding new energy automobile policy, subsiding development and utilization of clean energy policy, motor vehicle restriction policy and subsiding introduction of $PM_{2.5}$ treatment technology can save energy and improve environment better, while improving energy economic efficiency and environmental economic efficiency. Main conclusions can be drawn as follows.

1. Economic development

Restricted by the goal of reducing 30% haze pollutants emission of 2012 in 2030, the average annual economic growth rate of Hebei province from 2012 to 2030 should not exceed 5.87%, and the total economic output would reach 7.42 trillion RMB Yuan by 2030 (calculated at the same price in 2012). With the development of economy and society, efficient use of energy and environmental protection, Hebei province should vigorously develop green and environmental protection agriculture, maintain stable growth of primary industry output value, and reduce energy use and pollutant emissions of primary industry. The development of high energy-consuming and high polluting secondary industries such as the construction material industry, chemical, metallurgical and power industries should be restricted, while the development of other industries with less pollution and less energy consumption ought to be prompted so that the green and healthy grow of the secondary industry could be achieved. In addition, the tertiary industry should be vigorously developed to gradually increase the output value because it consumes less energy and has less pollution. According to the industrial development basis of Hebei province in 2012, the ratio of output value of three industries in Hebei province can be optimized to 13:37:50 in 2030.

2. Research conclusions on energy utilization

The structure of energy utilization has been optimized and the intensity of energy utilization has been greatly reduced. In 2030, coal, oil, natural gas, traditional electricity and clean energy accounts for 54.03%, 27.59%, 3.06%, 5.33% and 10% of the total energy utilization, respectively. In 2030, the intensity of energy utilization drops to 0.58 tons of standard coal/RMB 10,000 Yuan GRP, 49% less than that in 2012.

3. Research conclusions on environmental improvement

The total amount and intensity of haze pollutant emissions have been greatly reduced. Compared with 2016, the emission of SO_2, NOx, PM_{10}, CO, VOCs, $PM_{2.5}$ and GHG in Hebei province in 2030 can be reduced by 52%, 40%, 18%, 6%, 32% and -3.3% to 3400 tons, 2000 tons, 2700 tons, 14,300 tons, 2900 tons, 1700 tons and 19,500 tons, respectively. In addition, the emission intensity of SO_2, NOx, PM_{10}, CO, VOCs, $PM_{2.5}$ and GHG in Hebei province in 2030 could be reduced to 0.0029 tons/10,000

RMB Yuan, 0.0017 tons/10,000 RMB Yuan, 0.0024 tons/10,000 RMB Yuan, 0.00125 tons/10,000 RMB Yuan, 0.0026 tons/10,000 RMB Yuan, 0.0015 tons/10,000 RMB Yuan and 171 tons/10,000 RMB Yuan, respectively.

4. Policy effect conclusion

From the perspective of policy effect, the policies of subsiding for soil and water conservation, subsiding for development and utilization of clean energy, subsiding for new energy vehicles, motor vehicle restriction and subsiding for introduction of $PM_{2.5}$ treatment technology have diminishing effects. Comprehensive policy can better achieve the goal of sustainable development of economy, society, energy and environment than single policy, and the effect of "source governance" policy is better than that of "end governance" policy. From 2017 to 2021, the comprehensive policy has achieved remarkable simulation results in reducing emission, effectively promoting the rapid economic development. There are bottlenecks in the emission reduction effect of the comprehensive policies from 2022 to 2030. In order to achieve sustained and rapid economic growth under the constraint of emission reduction targets, new policies and technologies need to be further introduced.

5.2 Policy Recommendations

5.2.1 Recommendations on Policy Selection

Before 2021, Hebei province is supposed to adopt the policy combination of industrial restructuring, soil and water conservation, subsidizing new energy vehicles, subsidizing the development and utilization of clean energy, motor vehicle restriction and subsidizing the introduction of $PM_{2.5}$ treatment technology in order to achieve maximum economic development on the premise of meeting environmental requirements. To achieve the goal of sustainable and healthy development of economy, society, energy and environment faster and better, advanced policies and technologies need to be further introduced to maintain sustained and rapid economic growth under the constraints of environmental emission reduction targets since 2022.

5.2.2 Recommendations on Technical Selection

In order to realize the sustainable development of economy, society, energy and environment, Hebei province needs to invest in a number of ways and technologies for environmental improvement and energy utilization, such as new clean energy technology, $PM_{2.5}$ dust removal technology, enlarging forest area and the number of new energy vehicles (see Table 5.1). From 2012 to 2030, Hebei province should subsidize and invest 179082.21 MW of new clean energy installed capacity, including

Table 5.1 Environmental protection policy planning of Hebei province from 2012 to 2030

Year	New installed capacity of wind power (MW)	New photoelectric installation capacity (MW)	Number of electric bag compound dust removal plants for new units (Number)	New forest area (hectares)	New energy vehicles (Vehicles)
2012	0	0	0	0	0
2013	0	0	0	0	0
2014	0	0	0	0	0
2015	0	0	0	0	0
2016	0	0	2	0	0
2017	145428.56	0	4	67239	0
2018	3149.31	73.79	4	66230.41	0
2019	3063.95	217.79	4	65236.96	0
2020	2903.64	432.22	5	64258.4	0
2021	2669.17	716.41	0	63294.53	0
2022	2386.53	1046.44	0	62345.11	0
2023	2033.55	1442.76	0	61409.93	0
2024	1610.45	1905.21	0	60488.78	0
2025	1116.60	2434.40	0	59581.45	0
2026	552.60	2952.57	1	58687.73	0
2027	0.00	2946.26	0	57807.42	0
2028	0	0	0	56940.3	227585
2029	0	0	0	56086.2	218083
2030	0	0	0	55244.91	208979
Total	164914.36	14167.85	20	854851.13	654647

164914.36 MW of new wind power installed capacity and 14167.85 MW of new photovoltaic installed capacity. In addition, the government should subsidy $PM_{2.5}$ dust removal technology, added 20 units of electric bag composite dust removal plants. In addition, an additional 854,851.13 hectares of woodland, 654,647 new energy vehicles need to be invested.

5.2.3 *Recommendations on Financial Investment*

Specific subsidy schemes for various environmental policies are shown in Table 5.2. A total amount of RMB 376.209 billion is need for Hebei province to subsidize various environmental protection policies.

Table 5.2 Subsidies for environmental protection policy planning in Hebei province from 2017–2030 (Unit: 100 million RMB Yuan)

Year	Wind power subsidies	Photoelectric subsidies	Dust removal subsidy	Forest land subsidy	New energy vehicle subsidies
2017	295.68	0.00	1.63	2.69	0.00
2018	288.86	6.67	1.82	2.65	0.00
2019	275.58	19.72	2.10	2.61	0.00
2020	255.89	39.16	2.38	2.57	0.00
2021	232.01	64.96	0.50	2.53	0.00
2022	202.04	94.97	0.50	2.49	0.00
2023	165.97	131.06	0.52	2.46	0.00
2024	123.74	173.22	0.63	2.42	0.00
2025	75.39	221.52	0.71	2.38	0.00
2026	27.92	268.94	0.78	2.35	0.00
2027	27.92	269.05	0.71	2.31	0.00
2028	27.92	3.27	0.62	2.28	125.17
2029	27.92	3.27	0.55	2.24	119.95
2030	27.92	3.27	0.55	2.21	114.94
Total	2054.76	1299.08	14.00	34.19	360.06

Subsidy for the development and utilization of clean energy should be, 335.384 billion RMB Yuan, including 205.476 billion RMB Yuan for wind power generation and 129.908 billion RMB Yuan for solar power generation, which is 89.15% of the total financial subsidy. $PM_{2.5}$ dust removal technology (electric bag compound dust removal plant), soil and water conservation policy (forest land construction subsidy), and new energy vehicles will need 1.4 billion RMB Yuan, 3.419 billion RMB Yuan, 36.006 billion RMB Yuan, respectively, accounting for 0.37%, 0.91%, 9.57% of the financial investment.

Appendix

Appendix 1: Intermediate Input Coefficient of Industries in Hebei Province

Industry	Agriculture	Construction material industry	Chemical industry	Metallurgy	Power	Other industries	Retail and wholesale	Traffic and warehousing	Other services
Agriculture	0.1483	0.0002	0.0756	0.0000	0.0175	0.1310	0.0155	0.0102	0.0049
Construction material industry	0.0259	0.3294	0.0695	0.0267	0.0705	0.1050	0.0927	0.0186	0.0151
Chemical industry	0.0286	0.0004	0.3986	0.0233	0.0703	0.1522	0.0197	0.0399	0.0364
Metallurgy	0.0001	0.0009	0.0040	0.2742	0.0317	0.4196	0.0021	0.0400	0.0293
Power	0.0000	0.0001	0.0011	0.0032	0.3311	0.3927	0.0070	0.0299	0.0767
Other industries	0.0579	0.0055	0.0406	0.1445	0.0330	0.3608	0.0166	0.0304	0.0326
Retail and wholesale	0.0000	0.0002	0.0005	0.0000	0.0075	0.0125	0.0105	0.0317	0.1229
Traffic and warehousing	0.0001	0.0010	0.0166	0.0027	0.0201	0.2468	0.0026	0.1400	0.1119
Other services	0.0049	0.0008	0.0308	0.0003	0.0259	0.1254	0.0060	0.0269	0.2094

G. Zhang et al., *The Effect Evaluation of Haze Governance Policies in Hebei Province–Based on I-O Model*, https://doi.org/10.1007/978-981-15-3757-8

Appendix 2: Additional Value Ratio of Industries in Hebei Province

Agriculture	Construction material industry	Chemical industry	Metallurgy	Power	Other industries	Retail and wholesale	Traffic and warehousing	Other services
0.5967	0.2466	0.231	0.19796	0.1582	0.278159	0.814158	0.458215	0.558025

Appendix 3: Coefficient of Land Pollution in Hebei Province (t/ha)

Land type	SO_2	NO_x	PM_{10}	CO	VOCs	$PM_{2.5}$	GHG
Woodland	0.000000	0.000000	0.000000	0.000000	0.000000	0.000000	0.000000
Construction	0.000000	0.000000	0.089954	0.000000	0.000000	0.024935	47.810000
Other	0.101408	0.090691	0.139645	1.601338	0.213300	0.061076	33.640000

Appendix 4: Conversion Coefficients of $PM_{2.5}$ Treatment Technologies

Technical types	Investment conversion coefficient	Management conversion coefficient
Bag dust removal technology	0.020790	2.842276
Electric bag compound dust removal of unit	4.644976	0.013348
Wet electrostatic precipitation technology in coal-fired power plant	0.421080	0.296381
Biological nanofilm dust suppression technology	0.021780	0.726000

Appendix 5: Emission Coefficient of Primary and Tertiary Industries in Hebei Province (t/10,000 RMB Yuan)

Industry	SO_2	NOx	PM_{10}	CO	VOCs	$PM_{2.5}$	GHG
Agriculture	0.000182	0.000164	0.000031	0.000088	0.000127	0.000386	0.247724
Retail and wholesale	0.000190	0.000080	0.000039	0.000124	0.000057	0.000869	0.271118
Traffic and warehousing	0.000351	0.000443	0.000050	0.000128	0.000350	0.000104	0.487218
Other services	0.000179	0.000071	0.000037	0.000118	0.000050	0.000839	0.234947

Appendix 6: Emission Coefficient of Industrial Combustion Process in Hebei Province (t/10,000 RMB Yuan)

Industry	SO_2	NOx	PM_{10}	CO	VOCs	$PM_{2.5}$	GHG
Construction material industry	0.050293	0.021315	0.012798	0.004854	0.002260	0.002223	4.191630
Chemical industry	0.000288	0.000091	0.000249	0.000672	0.000148	0.000157	2.148470
Metallurgy	0.002669	0.011870	0.000927	0.001738	0.000975	0.000896	0.226167
Power	0.000452	0.000086	0.000060	0.001729	0.001699	0.000148	0.638973
Other industries	0.000058	0.000029	0.000023	0.000018	0.000012	0.000008	0.637227

Appendix 7: Discharge Coefficient of Industrial in Hebei Province (t/10,000 RMB Yuan)

Industry	SO_2	NOx	PM_{10}	CO	VOCs	$PM_{2.5}$	GHG
Construction material industry	0.000579	0.000244	0.000719	0.000492	0.000445	0.000364	0.000000
Chemical industry	0.000579	0.000244	0.000719	0.000492	0.000445	0.000353	0.000000
Metallurgy	0.000579	0.000244	0.000719	0.000492	0.000445	0.000353	0.000000
Power	0.000579	0.000244	0.000719	0.000492	0.000445	0.000353	0.000000
Other industries	0.000579	0.000244	0.000719	0.000492	0.000445	0.000353	0.000000

Appendix 8: Final Consumption Energy Demand Coefficient of Hebei Province (Ton of Standard Coal/10000 RMB Yuan)

Industry	Coal	Petroleum	Natural gas	Power
Agriculture	0.214334	0.018585	0.004680	0.003768
Construction material industry	0.712409	0.061774	0.015556	0.012524
Chemical industry	0.283358	0.024570	0.006187	0.004981
Metallurgy	0.000000	0.000000	0.000000	0.000000
Power	0.048776	0.004229	0.001065	0.000857
Other industries	0.214334	0.018585	0.004680	0.003768
Retail and wholesale	0.214334	0.018585	0.004680	0.003768
Traffic and warehousing	0.214334	0.018585	0.004680	0.003768
Other services	0.214334	0.018585	0.004680	0.003768

Appendix 9: Per Capita Consumption Coefficient of Hebei Province (10,000 RMB Yuan/People)

Agriculture	Construction material industry	Chemical industry	Metallurgy	Power	Other industries	Retail and wholesale	Traffic and warehousing	Other services
0.13277	0.0390	0	0.00888	0.0295	0.34767	0.12091	0.03037	0.81136

Appendix 10: Emission Coefficient of Motor Vehicles in Hebei Province (Tons/Vehicle)

Type	SO_2	NO_x	PM_{10}	CO	VOC_s	$PM_{2.5}$	GHG
Car	0.000100	0.001000	0.000200	0.014000	0.015000	0.000010	3.223000
Tractor	0.000050	0.000500	0.000040	0.000100	0.000025	0.000756	0.119100
Motorcycle	0.000030	0.001000	0.000100	0.047000	0.008000	0.000674	0.805000

Appendix 11: Energy Demand Coefficient of Industries in Hebei Province (Ton of Standard Coal/10000 RMB Yuan)

Industry	Coal	Petroleum	Natural gas	Power
Agriculture	0.036763	0.067711	0.001802	0.026205
Construction material industry	0.081886	0.000000	0.005118	0.015354
Chemical industry	0.221788	0.006344	0.021629	0.000000
Metallurgy	0.791849	0.016243	0.004061	0.000000
Power	1.575583	0.000000	0.000000	0.000000
Other industries	0.225094	0.011131	0.008754	0.002377
Retail and wholesale	0.084368	0.010124	0.015186	0.059058
Traffic and warehousing	0.007644	0.214027	0.001390	0.008570
Other services	0.081463	0.007064	0.001779	0.001432

Appendix 12: The Final Consumption Emission Factor of Hebei Province (Ton/10000 RMB Yuan)

Industry	SO_2	NOx	PM_{10}	CO	VOCs	$PM_{2.5}$	GHG
Agriculture	0.000470	0.000188	0.000097	0.000310	0.000132	0.002208	0.618161
Construction material industry	0.000107	0.000043	0.000022	0.000071	0.000030	0.000502	0.140675
Chemical industry	0.000000	0.000000	0.000000	0.000000	0.000000	0.000000	0.000000
Metallurgy	0.001563	0.000625	0.000323	0.001031	0.000439	0.007338	2.054662
Power	0.000622	0.000249	0.000128	0.000410	0.000175	0.002919	0.817234
Other industries	0.000470	0.000188	0.000097	0.000310	0.000132	0.002208	0.618161
Retail and wholesale	0.000470	0.000188	0.000097	0.000310	0.000132	0.002208	0.618161
Traffic and warehousing	0.000470	0.000188	0.000097	0.000310	0.000132	0.002208	0.618161
Other services	0.000470	0.000188	0.000097	0.000310	0.000132	0.002208	0.618161

Appendix 13: Depreciation Rate of Industrial Capital in Hebei Province

Agriculture	Construction material industry	Chemical industry	Metallurgy	Power	Other industries	Retail and wholesale	Traffic and warehousing	Other services
0.1	0.1	0.1	0.1	0.1	0.1	0.1	0.1	0.1

Appendix 14: Capital Input-Output Ratio of Hebei Province

Agriculture	Construction material industry	Chemical industry	Metallurgy	Power	Other industries	Retail and wholesale	Traffic and warehousing	Other services
0.07966	0.1483	0.00406	0.00017	0	0.393281	0.139481	0.011545	0.034145

Appendix 15: Conversion Coefficient of Clean Energy Technology

Type	Investment conversion coefficient	Management conversion coefficient
Wind power technology 1	0.276100	0.007070
Wind power technology 2	0.263300	0.006590
Photoelectric technology 1	0.286100	0.008950
Photoelectric technology 2	0.271000	0.008130

Bibliography

Akimoto, H. (2003). Global air quality and pollution. *Science, 302,* 1716–1719.

Asian Development Bank. (2012). Toward an environmentally sustainable future: Country environmental analysis of the People's Republic of China. Asian Development Bank.

Bench, G., & Herckes, P. (2004). Measurement of contemporary and fossil carbon contents of $PM_{2.5}$ aerosols: Results from Turtleback Dome, Yosemite National Park. *Environmental Science & Technology, 38*(8), 2424–2427.

Callen, M. S., Iturmendi, A., & Lopez, J. M. (2014). Source apportionment of atmospheric $PM_{2.5}$-bound polycyclic aromatic hydrocarbons by a PMF receptor model. Assessment of potential risk for human health. *Environmental Pollution, 195,* 167–177.

Cesari, D., Genga, A., Ielpo, P., Siciliano, M., Mascolo, G., Grasso, F. M., et al. (2014). Source apportionment of $PM_{2.5}$ in the harbour-industrial area of Brindisi (Italy): Identification and estimation of the contribution of in-port ship emissions. *Science of the Total Environment, 497,* 392–400.

Chen, H. S., Ge, B. Z., Yang, W. Y., Wang, Z. F., Huang, S., Wang, Y. L., et al. (2015). Modeling study of source contributions and emergency control effects during a severe haze episode over the Beijing-Tianjin-Hebei area. *Science China-Chemistry, 58*(9), 1403–1415.

Chen, S. M., & He, L. Y. (2014). Welfare loss of China's air pollution: How to make personal vehicle transportation policy. *China Economic Review, 31,* 106–118.

Chiesa, M., Perrone, M. G., Cusumano, N., Ferrero, L., Sangiorgi, G., Bolzacchini, E., et al. (2014). An environmental, economical and socio-political analysis of a variety of urban air-pollution reduction policies for primary PM_{10} and NOx: The case study of the Province of Milan (Northern Italy). *Environmental Science & Policy, 44,* 39–50.

China Academy of Sciences Special Overall Group on "Tracing and Control of Atmospheric Haze". (2005). Beijing-Tianjin-Hebei $PM_{2.5}$ control effect evaluation report since the implementation of the "Atmosphere Ten". *Bulletin of Chinese Academy of Sciences, 30*(5), 668–678 (in Chinese).

Choi, J. K., Heo, J. B., Ban, S. J., Yi, S. M., & Zoh, K. D. (2013). Source apportionment of $PM_{2.5}$ at the coastal area in Korea. *Science of the Total Environment, 447,* 370–380.

Chun, S. L., Feng, K. D., Ke, B. H., & Yong, L. M. (2016). Review on recent progress in observations, source identifications and countermeasures of $PM_{2.5}$. *Environment International, 86,* 150–170.

Chung, W., Tohno, S., & Shim, S. (2009). An estimation of energy and GHG emission intensity caused by energy consumption in Korea: An energy IO approach. *Applied Energy, 86*(10), 1902–1941.

Cohan, D. S., Boylan, J. W., Marmur, A., & Khan, M. N. (2007). An integrated framework for multipollutant air quality management and its application in Georgia. *Environmental Management, 40*(4), 545–554.

G. Zhang et al., *The Effect Evaluation of Haze Governance Policies in Hebei Province–Based on I-O Model*, https://doi.org/10.1007/978-981-15-3757-8

Du, Z. Y., He, K. B., Cheng, Y., Duan, F. K., Ma, Y. L., Liu, J. M., et al. (2014). A yearlong study of water-soluble organic carbon in Beijing I: Sources and its primary vs. secondary nature. *Atmospheric Environment, 92*, 514–521.

Forsund, F. R. (1985). Input-output models: National economic models and the environment. In A. V. Kneese & J. L. Sweeney (Eds.), *Handbook of national resource and energy economics* (Vol. 1, pp. 325–341).

Gao, J. J., Tian, H. Z., Cheng, K., Lu, L., Wang, Y. X., Wu, Y., et al. (2014). Seasonal and spatial variation of trace elements in multi-size airborne particulate matters of Beijing, China: Mass concentration, enrichment characteristics, source apportionment, chemical speciation and bioavailability. *Atmospheric Environment, 99,* 257–265.

Gao, J. L., Yuan, Z. W., Liu, X. W., Xia, X. M., Huang, X. J., & Dong, Z. F. (2015). Improving air pollution control policy in China—A perspective based on cost-benefit analysis. *Science of the Total Environment, 543,* 307–314.

Gao, M., Guttikunda, S. K., Carmichael, G. R., Wang, Y. S., Liu, Z. R., Stanier, C., et al. (2015). Health impacts and economic losses assessment of the 2013 severe haze event in Beijing area. *Science of the Total Environment, 511,* 553–561.

Guo, S., Hu, M., Zamora, M. L., Peng, J. F., Shang, D. J., Zheng, J., et al. (2014). Elucidating severe urban haze formation in China. *Proceedings of the National Academy of Sciences of the United States of America, 111*(49), 17373–17378.

Han, L. J., Zhou, W. Q., Li, W. F., & Li, L. (2014). Impact of urbanization level on urban air quality: A case of fine particles ($PM_{2.5}$) in Chinese cities. *Environmental Pollution, 194*, 163–170.

Hanna, B., Nicole, A. H. J., Paul, H. F., Gerard, P. A. K., Ernie, P. W., Flemming, R. C., et al. (2012). Impact of low emission zones and local traffic policies on ambient air pollution concentrations. *Science of the Total Environment, 435–436,* 132–140.

Hannon, B. (2001). Ecological pricing and economic efficiency. *Ecological Economics, 36*, 19–30.

Hasheminassab, S., Daher, N., Ostro, B. D., & Sioutas, C. (2014). Long-term source apportionment of ambient fine particulate matter ($PM_{2.5}$) in the Los Angeles Basin: A focus on emissions reduction from vehicular sources. *Environmental Pollution, 193*, 54–64.

He, H., & Yu, Y. B. (2005). Selective catalytic reduction of NOx over Ag/Al2O3 catalyst: From reaction mechanism to diesel engine test. *Catalysis Today, 100*(1–2), 37–47.

Higano, Y., & Sawada, T. (1997). The dynamic policy to improve the water quality of lake Kasumigaura. *Studies in Regional Science, 26*(1), 75–86.

Higano, Y., & Yoneta, A. (1999). Economic policies to relieve contamination of Lake Kasumigaura. *Studies in Regional Science, 29*(3), 205–218.

Hirose, F., & Higano, Y. (2000). A simulation analysis to reduce pollutants from the catchment area of lake Kasumigaura. *Studies in Regional science, 30*(1), 47–63.

Hixson, M., Mahmud, A., Hu, J. L., Bai, S., Niemeier, D. A., Handy, S. L., et al. (2009). Influence of regional development policies and clean technology adoption on future air pollution exposure. *Atmospheric Environment, 44*(4), 552–562.

Hristu, V. D., Karagianni, S., Pempetzoglouc, M., et al. (2010). Optimizing product with energy and GHG emission constraints in Greece: An input-output analysis. *Energy Policy, 38*(2010), 1566–1577.

Hu, J. L., Wang, Y. G., Ying, Q., & Zhang, H. L. (2014). Spatial and temporal variability of $PM_{2.5}$ and PM_{10} over the North China plain and the Yangtze River Delta, China. *Atmospheric Environment, 95*, 598–609.

Huang, B., Liu, M., Ren, Z. F., Bi, X. H., Zhang, G. H., Sheng, G. Y., et al. (2013). Chemical composition, diurnal variation and sources of $PM_{2.5}$ at two industrial sites of South China. *Atmospheric Pollution Research, 4*(3), 298–305.

Huang, K., Zhang, X. Y., & Lin, Y. F. (2015). The "APEC Blue" phenomenon: Regional emission control effects observed from space. *Atmospheric Research, 164,* 65–75.

Huang, R. J., Zhang, Y. L., Bozzetti, C., Ho, K. F., Cao, J. J., Han, Y. M., et al. (2014). High secondary aerosol contribution to particulate pollution during haze events in China. *Nature, 514,* 218–222.

Jiang, X. J., Hong, C. P., Zheng, Y. X., Zheng, B., Guan, D. B., Gouldson, A., et al. (2015). What extent can China's near-term air pollution control policy protect air quality and human health? A case study of the Pearl River Delta region. *Environmental Research Letters, 10*(10), 1–13.
Kanada, M., Fujita, T., Fujii, M., & Ohnishi, S. (2013). The long-term impacts of air pollution control policy: Historical links between municipal actions and industrial energy efficiency in Kawasaki City, Japan. *Journal of Cleaner Production, 58*(s), 92–101.
Kulkarni, P., Chellam, S., & Fraser, M. P. (2007). Tracking petroleum refinery emission events using lanthanum and lanthanides as elemental markers for $PM_{2.5}$. *Environmental Science & Technology, 41*(19), 6748–6754.
Lee, K. S. (1982). A generalized input-output model of an economy with environmental protection. *The Review of Economics and Statistics, 64*(3), 466–473.
Lenzen, M. (1998). Primary energy and GHG embodied in Australian final consumption: An input-output analysis. *Energy Policy, 26*(6), 495–506.
Leontief, W. W. (1977). *The future of the world economy*. New York: Oxford University Press.
Leontief, W. W. (1986). *Input-output economics* (2nd ed.). New York: Oxford University Press.
Li, C. H., & Zhao, J. Y. (2017). Indirect economic loss estimation of inter-industrial sectors: Triggered from agricultural economic loss induced from Haze in Beijing. *Chinese Agricultural Science Bulletin, 33*(10), 118–124. (in Chinese).
Li, X., Zhang, Q., Zhang, Y., Zheng, B., Wang, K., Chen, Y., et al. (2015). Source contributions of urban $PM_{2.5}$ in the Beijinge-Tianjine-Hebei region: Changes between 2006 and 2013 and relative impacts of emissions and meteorology. *Atmospheric Environment, 123*(A), 229–239.
Liu, T., Zhang, Y. H., Xu, Y. J., Lin, H. L., Xu, X. J., Luo, Y., et al. (2014). The effects of dust-haze on mortality are modified by seasons and individual characteristics in Guangzhou, China. *Environmental Pollution, 187*, 116–123.
Liu, Y. C., He, H., & Mu, Y. J. (2008). Heterogeneous reactivity of carbonyl sulfide on α-Al2O3 and γ-Al2O3. *Atmospheric Environment, 45*(5), 960–969.
Ma, G. X., Yu, F., Zhang, Y. S., et al. (2019). The effect assessment of implementation of the action plan on prevention and control of air pollution and its impact on life expectancy per capita in China. *Research of Environmental Sciences*. https://doi.org/10.13198/j.issn.1001-6929.2019.07.22 (in Chinese).
Malla, S. (2014). Assessment of mobility and its impact on energy use and air pollution in Nepal. *Energy, 69*, 485–496.
Mao, X. Q., Zhou, J., & Corsetti, G. (2014). How well have China's recent five-year plans been implemented for energy conservation and air pollution control? *Environmental Science & Technology, 48*(17), 10036–10044.
Mizunoya, T., Sakurai, K., Kobayashi, S., et al. (2007). A simulation analysis of synthetic environment policy: Effective utilization of biomass resources and reduction of environmental burdens in Kasumigaura basin. *Studies in Regional Science, 36*(2), 355–374.
Moffatt, I., & Hanley, N. (2001). Modeling sustainable development: Systems dynamic and input-output approaches. *Environmental Modeling and Software, 16*, 545–557.
Mueller, S. F., & Mallard, J. W. (2011). Contributions of natural emissions to ozone and $PM_{2.5}$ as simulated by the Community Multiscale Air Quality (CMAQ) model. *Environmental Science & Technology, 45*(11), 4817–4823.
Muller, N. Z. (2012). The design of optimal climate policy with air pollution co-benefits. *Resource and Energy Economics, 34*(4), 696–722.
Novan, K. (2015). Valuing the wind: Renewable energy policies and air pollution avoided. *American Economic Journal-Economic Policy, 7*(3), 291–326.
Oliveira, C., & Antunes, H. (2004). A multiple objective model to deal with economy-energy-environment interactions. *European Journal of Operational Research, 153*(2), 370–385.
Perrings, C. (1987). *Economy and environment*. New York: Cambridge University Press.
Qiu, X. Z., Zhu, Y., Jang, C., Lin, C. J., Wang, S. X., Fu, J., et al. (2015). Development of an integrated policy making tool for assessing air quality and human health benefits of air pollution control. *Frontiers of Environmental Science & Engineering, 9*(6), 1056–1065.

Quan, J., Zhang, Q., He, H., Liu, J., Huang, M., & Jin, H. (2011). Analysis of the formation of fog and haze in North China Plain (NCP). *Atmospheric Chemistry and Physics, 11*, 8205–8214.

Rafaj, P., & Kypreos, S. (2007). Internalisation of external cost in the power generation sector: Analysis with global multi-regional MARKAL model. *Energy Policy, 35*(2), 828–843.

Rao, S., Pachauri, S., Dentener, F., Kinney, P., Klimont, Z., Riahi, K., et al. (2013). Better air for better health: Forging synergies in policies for energy access, climate change and air pollution. *Global Environmental Change, 23*(5), 1122–1130.

Rive, N. (2010). Climate policy in Western Europe and avoided costs of air pollution control. *Economic Modeling, 27*(1), 103–115.

Ru, J. H., Zhang, Y. L., Carlo, B., Ho, K. F., Cao, J. J., Han, Y. M., et al. (2014). High secondary aerosol contribution to particulate pollution during haze events in China. *Nature, 514*, 218–222.

Shan, W. P., Liu, F. D., He, H., Shi, X. Y., & Zhang, C. B. (2011). The remarkable improvement of a Ce–Ti based catalyst for NOx abatement, prepared by a homogeneous precipitation method. *ChemCatChem, 3*(8), 1286–1289.

Shen, G. F. (2015). Quantification of emission reduction potentials of primary air pollutants from residential solid fuel combustion by adopting cleaner fuels in China. *Journal of Environmental Sciences, 37*(1), 1–7.

Shen, Z. X., Cao, J. J., Arimoto, R., Han, Z. W., Zhang, R. J., Han, Y. M., et al. (2009). Ionic composition of TSP and $PM_{2.5}$ during dust storms and air pollution episodes at Xi'an, China. *Atmospheric Environment, 43*(18), 2911–2918.

Si, J. P., Liu, X. W., Xu, M. H., Sheng, L., Zhou, Z. J., Wang, C., et al. (2014). Effect of kaolin additive on $PM_{2.5}$ reduction during pulverized coal combustion: Importance of sodium and its occurrence in coal. *Applied Energy, 114*(s), 434–444.

Song, C., Pei, T., & Yao, L. (2015a). Analysis of the characteristics and evolution modes of $PM_{2.5}$ pollution episodes in Beijing, China during 2013. *International Journal of Environmental Research and Public Health, 12*(2), 1099–1111.

Song, J. N., Yang, W., Higano, Y., & Wang, X. E. (2015b). Introducing renewable energy and industrial restructuring to reduce GHG emission: Application of a dynamic simulation model. *Energy Conversion and Management, 96*, 625–636.

Song, J. N., Yang, W., Higano, Y., & Wang, X. E. (2015c). Modeling the development and utilization of bioenergy and exploring the environmental economic benefits. *Energy Conversion and Management, 103*, 836–846.

Song, J. N., Yang, W., Higano, Y., & Wang, X. E. (2015d). Dynamic integrated assessment of bioenergy technologies for energy production utilizing agricultural residues: An input-output approach. *Applied Energy, 158*, 178–189.

Sueyoshi, T., & Yuan, Y. (2015). China's regional sustainability and diversified resource allocation: DEA environmental assessment on economic development and air pollution. *Energy Economics, 49*, 239–256.

Sun, H. X., & Li, S. (2018). Spatial coupling relationship between atmospheric haze and coal consumption and environmental taxation: A case study of 31 provinces and cities in China. *Inquiry into Economic Issues, 1*, 155–166. (in Chinese).

Sun, Y., Wang, L. L., Wang, Y. S., Liu, Q., & Liu, Z. R. (2011). In situ measurements of SO_2, NOx, NOy, and O3 in Beijing, China during August 2008. *Science of the Total Environment, 409*(5), 933–940.

Sun, Y. L., Zhuang, G. S., Tang, A. H., Wang, Y., & An, Z. S. (2006). Chemical characteristics of $PM_{2.5}$ and PM_{10} in haze-fog episodes in Beijing. *Environmental Science & Technology, 40*(10), 3148–3155.

Tan, J. H., Duan, J. C., He, K. B., Ma, Y., Duan, F. K., Chen, Y., et al. (2009). Chemical characteristics of $PM_{2.5}$ during a typical haze episode in Guangzhou. *Journal of Environmental Sciences, 21*(6), 774–781.

Tan, M., Jiang, M. Y., & Geng, A. Q. (2018). Approach to the correlation between haze occurrence and urban rural planning in urban agglomeration areas. *China Population, Resources and Environment, 28*(7), 67–70. (in Chinese).

Turner, R. K. (2000). Integrating natural and socio-economic science in coastal management. *Journal of Marine System, 25*, 447–460.

Vafa-Arani, H., Jahani, S., Dashti, H., Heydari, J., & Moazen, S. (2014). A system dynamics modeling for urban air pollution: A case study of Tehran, Iran. *Transportation Research Part D: Transport and Environment, 31*, 21–36.

Vedrenne, M., Borge, R., Lumbreras, J., Conlan, B., Rodriguez, M. E., de Andres, J. M., et al. (2015). An integrated assessment of two decades of air pollution policy making in Spain: Impacts, costs and improvements. *Science of the Total Environment, 527*, 351–361.

Wang, G. Z., Gu, S. J., & Chen, J. B. (2016). Assessment of the indirect economic loss caused by heavy haze in Beijing based on input-output model. *Environmental Engineering, 1*, 121–125 (in Chinese).

Wang, F. W., Lin, T., Feng, J. L., Fu, H. Y., & Guo, Z. G. (2015a). Source apportionment of polycyclic aromatic hydrocarbons in $PM_{2.5}$ using positive matrix factorization modeling in Shanghai, China. *Environmental Science & Technology, 17*(1), 197–205.

Wang, H. L., Qiao, L. P., Lou, S. R., Zhou, M., Ding, A. J., Huang, H. Y., et al. (2015b). Chemical composition of $PM_{2.5}$ and meteorological impact among three years in urban Shanghai, China. *Journal of Cleaner Production, 112*, 1302–1311.

Wang, L. T., Wei, Z., Yang, J., Zhang, Y., Zhang, F. F., Su, J., et al. (2014). The 2013 severe haze over southern Hebei, China: Model evaluation, source apportionment, and policy implications. *Atmospheric Chemistry and Physics, 14*(6), 3151–3173.

Wang, K., & Liu, Y. N. (2014). Can Beijing fight with haze? Lessons can be learned from London and Los Angeles. *Natural Hazards, 72*, 1265–1274.

Wang, P., Cao, J. J., Shen, Z. X., Han, Y. M., Lee, S. C., Huang, Y., et al. (2015c). Spatial and seasonal variations of $PM_{2.5}$ mass and species during 2010 in Xi'an, China. *Science of the Total Environment, 508*, 477–487.

Wang, Q. Z., Zhuang, G. S., Huang, K., Liu, T. N., Deng, C. R., Xu, J., et al. (2015). Probing the severe haze pollution in three typical regions of China: Characteristics, sources and regional impacts. *Atmospheric Environment, 120*, 76–88.

Wang, S. X., Xing, J., Zhao, B., Jang, C., & Hao, J. M. (2014). Effectiveness of national air pollution control policies on the air quality in metropolitan areas of China. *Journal of Environmental Sciences, 26*(1), 13–32.

Wei, Y. J., Wang, Q. Y., Zhang, L., Awaya, I., Ji, M. J., Li, H. X., et al. (2013). Effect of magnesium additives on $PM_{2.5}$ reduction during pulverized coal combustion. *Fuel Processing Technology, 105*(S), 188–194.

WHO. (2014). Ambient (outdoor) air quality and health. http://www.who.int/mediacentre/factsheets/fs313/en/.

Word Bank. (2007). Cost of pollution in the People's Republic of China: Economic estimates of physical damages. In Conference Edition. Washington, DC.

Wu, S. W., Deng, F. R., Wei, H. Y., Huang, J., Wang, X., Hao, Y., et al. (2014). Association of cardiopulmonary health effects with source-appointed ambient fine particulate in Beijing, China: A combined analysis from the Healthy Volunteer Natural Relocation (HVNR) study. *Environmental Science & Technology, 48*(6), 3438–3448.

Xiang, N., Sha, J. H., Yan, J. J., & Xu, F. (2014). Dynamic modeling and simulation of water environment management with a focus on water recycling. *Water, 6*(1), 17–31.

Xiang, N., Xu, F., Shi, M. J., & Zhou, D. Y. (2015). Assessing the potential of using water reclamation to improve the water environment and economy: Scenario analysis of Tianjin, China. *Water Policy, 17*(3), 391–408.

Xu, F., Xiang, N., Nijkamp, P., & Higano, Y. (2013). Dynamic simulation of China's carbon intensity and energy intensity evaluation focusing on industry and energy structure adjustment by 2020. *Environmental Engineering and Management Journal, 12*(10), 1897–1901.

Xu, F., Xiang, N., & Higano, Y. (2015). Comprehensive evaluation of environmental policies for sustainable development in Jaxing city, China. *Environmental Engineering and Management Journal, 14*(5), 1079–1088.

Yan, R. C., Yu, S. C., Zhang, Q. Y., Li, P. F., Wang, S., Chen, B. X., et al. (2015). A heavy haze episode in Beijing in February of 2014: Characteristics, origins and implications. *Atmospheric Pollution Research, 6*, 867–876.

Yan, J. J., Sha, J. H., Chu, X., Xu, F., & Higano, Y. (2014). Endogenous derivation of optimal environmental policies for proper treatment of stockbreeding wastes in the upstream region of the Miyun Reservoir, Beijing. *Papers in Regional Science, 93*(2), 477–500.

Yan, J. J., Sha, J. H., Chu, X., Xu, F., & Xiang, N. (2014b) Dynamic evaluation of water quality improvement based on effective utilization of stockbreeding biomass resource. *Sustainability, 6*, 8218–8236.

Yang, F., Tan, J., Zhao, Q., Du, Z., He, K., Ma, Y., et al. (2011). Characteristics of $PM_{2.5}$ speciation in representative megacities and across China. *Atmospheric Chemistry and Physics, 11*(11), 5207–5219.

Yang, W., Song, J. N., Higano, Y., & Tang, J. (2015a). An integrated simulation model for dynamically exploring the optimal solution to mitigating water scarcity and pollution. *Sustainability, 7*, 1774–1797.

Yang, W., Song, J. N., Higano, Y., & Tang, J. (2015b). Exploration and assessment of optimal policy combination for total water pollution control with a dynamic simulation model. *Journal of Cleaner Production, 102*, 342–352.

Yu, S. C. (2014). Water spray geoengineering to clean air pollution for mitigating haze in China's cities. *Environmental Chemistry Letters, 12*(1), 109–116.

Yuan, X. L., Mu, R. M., Zuo, J., & Wang, Q. S. (2015). Economic development, energy consumption, and air pollution: A critical assessment in China. *Human and Ecological Risk Assessment, 21*(3), 781–798.

Zhang, D. S., Aunan, K., Seip, H. M., Larssen, S., Liu, J. H., & Zhang, D. S. (2010). The assessment of health damage caused by air pollution and its implication for policy making in Tai yuan, Shanxi, China. *Energy Policy, 38*(1), 491–502.

Zhang, G. F. (2014). Dynamic simulation of environmental and economic impacts of sewage sludge treatment in Beijing. China University of Geosciences (Beijing) (in Chinese).

Zhang, G. F., Sha, J. H., Wang, T. Y., Yan, J. J., & Higano, Y. (2013). Comprehensive evaluation of socio-economic and environmental impacts using membrane bioreactors for sewage treatment in Beijing. *Journal of Pure and Applied Microbiology*, *7*(S), 553–564.

Zhang, Q. Y., Yan, R. C., Fan, J. W., Yu, S. C., Yang, W. D., Li, P. F., et al. (2015). A heavy haze episode in Shanghai in December of 2013: Characteristics, origins and implications. *Aerosol and Air Quality Research, 15*(5), 1881–1893.

Zheng, G. J., & Yang, L. K. (2015). Solutions of fog and haze weather from the perspective of economic development. *Ecological Economy, 31*(9), 34–38. (in Chinese).

Zheng, S. M., Yi, H. T., & Li, H. (2015). The impacts of provincial energy and environmental policies on air pollution control in China. *Renewable and Sustainable Energy Reviews, 49*, 386–394.

Zhou, Q., Yabar, H., Mizunoya, T., & Higano, Y. (2016). Exploring the potential of introducing technology innovation regulations in the energy sector in China: A regional dynamic evaluation model. *Journal of Cleaner Production, 112*, 537–1548.

Zhou, X. P., Xu, Y. Y., Yuan, S., Wu, C., & Zhang, H. (2015). Performance and potential of solar updraft tower used as an effective measure to alleviate Chinese urban haze problem. *Renewable & Sustainable Energy Reviews, 51*, 1499–1508.

Zhuang, X. L., Wang, Y. S., He, H., Liu, J. G., Wang, X. M., Zhu, T. Y., et al. (2014). Haze insights and mitigation in China: An overview. *Journal of Environmental Sciences, 26*, 2–12.

The manufacturer's authorised representative in the EU is Springer Nature Customer Service Centre GmbH, Europaplatz 3, 69115 Heidelberg, Germany. If you have any concerns regarding our products, please contact ProductSafety@springernature.com

Printed and bound by CPI Group (UK) Ltd, Croydon, CR0 4YY
25/09/2026
02222901-0014